Printed in the United States
by Baker & Taylor Publisher Services

SpringerBriefs in Applied Sciences and Technology

SpringerBriefs in Computational Intelligence

Series Editor

Janusz Kacprzyk, Systems Research Institute, Polish Academy of Sciences,
Warsaw, Poland

SpringerBriefs in Computational Intelligence are a series of slim high-quality publications encompassing the entire spectrum of Computational Intelligence. Featuring compact volumes of 50 to 125 pages (approximately 20,000–45,000 words), Briefs are shorter than a conventional book but longer than a journal article. Thus Briefs serve as timely, concise tools for students, researchers, and professionals.

More information about this subseries at http://www.springer.com/series/10618

Patricia Melin · Ivette Miramontes ·
German Prado Arechiga

Nature-inspired Optimization of Type-2 Fuzzy Neural Hybrid Models for Classification in Medical Diagnosis

 Springer

Patricia Melin
Division of Graduate Studies
Tijuana Institute of Technology
Tijuana, Baja California, Mexico

Ivette Miramontes
Division of Graduate Studies
Tijuana Institute of Technology
Tijuana, Baja California, Mexico

German Prado Arechiga
Cardiologia, Cardio Diagnostico
Tijuana, Baja California, Mexico

ISSN 2191-530X   ISSN 2191-5318 (electronic)
SpringerBriefs in Applied Sciences and Technology
ISSN 2625-3704   ISSN 2625-3712 (electronic)
SpringerBriefs in Computational Intelligence
ISBN 978-3-030-82218-7   ISBN 978-3-030-82219-4 (eBook)
https://doi.org/10.1007/978-3-030-82219-4

This Springer imprint is published by the registered company Springer Nature Switzerland AG
The registered company address is: Gewerbestrasse 11, 6330 Cham, Switzerland

Preface

Nowadays, soft computing techniques are used to solve different problems with the aim of reducing people's workload and obtaining better precision in the results. These techniques are used with great success to determine the risk diagnosis in different application areas in order to forecast any contingency situation in a given time and thus be able to act in time.

One of the interesting areas for the application of soft computing systems is the medical area. Therefore, this book proposes a new neuro-fuzzy hybrid model to provide a precise and timely medical diagnosis based on the patient's blood pressure, where it is proposed to use artificial neural networks, which perform the learning of different behaviors based on patient information, such as risk factors and blood pressure readings. In addition, to the use of fuzzy inference systems, which help to provide the classifications of the health status of the patient based on the results of the neural networks.

To reduce errors in this type of application, it is very important to use optimization algorithms, which help us find those parameters with which the best results are obtained, that is why to achieve precision in the proposed model, each one of the modules proposed in the model is optimized using different bio-inspired algorithms. This is done in order to make a comparison of the obtained results and determine with which of these methods the best result is obtained when providing the corresponding diagnosis.

Another way to improve the obtained results through the bio-inspired algorithms is to make some modifications to the algorithms. In this case, the algorithm that provides the worst error in the results is taken and a dynamic parameter adjustment is made using fuzzy systems, which was tested with different study cases to test and determine if there is an improvement in the results obtained.

This book is intended to be a reference for scientists and engineers interested in using different applied intelligent computing techniques in solving medical problems. Similarly, this book can serve as a reference for new proposals or to improve the results presented.

In Chap. 1, an introduction is presented where the current health problems that we face are discussed in general and we mention the importance of a timely diagnosis, for which, soft computing techniques are of great help. The way in which both artificial neural networks and fuzzy systems can be used to provide the diagnosis of the development of hypertension is discussed. In addition, antecedents of other research works are presented, where they diagnose different conditions with soft computing techniques.

To better understand this research, Chap. 2 describes different medical concepts, such as blood pressure and hypertension, to name a few. Also, the concepts of the different computing techniques used are described, such as artificial neural networks and fuzzy systems, in addition to explaining the different bio-inspired algorithms used in this book.

In Chap. 3, a detailed explanation of how the problem is to be solved is given, besides, an explanation is also given of how soft computing techniques were applied to the proposed model.

In Chap. 4, the different study cases carried out are explained, which cover the optimization of the different soft computing techniques used to provide the diagnosis based on the blood pressure of different patients.

The conclusions obtained from the experimentation are explained in Chap. 5, which corresponds to the development of the neuro-fuzzy hybrid model, to which we provide knowledge for decision-making and provide the corresponding medical diagnosis. With this model, a precise and timely diagnosis of the risk that a patient has in developing hypertension and a cardiovascular event is provided in a period of time. The proposed model presents efficient results when diagnosing a group of patients, where both their blood pressure and their risk factors vary significantly.

So far, the performance of the proposed model has presented good performance, but it is necessary to experiment with other scenarios, such as persons with severe hypertension. This requires obtaining more blood pressure readings from patients.

We end this preface of the book by giving thanks to all the people who have helped in this project and during the writing of this book. We would like to thank our funding institutions CONACYT and TNM of our country for their support within this project. We have to thank our institution, the Tijuana Institute of Technology, for always supporting our projects. Finally, we thank our families for their support and patience in this project.

Tijuana, Mexico   Patricia Melin
                                                                          Ivette Miramontes
                                                                     German Prado Arechiga

Contents

Chapter 1
Introduction to Soft Computing Applied in Medicine

Currently, hypertension represents a huge public health problem worldwide. Investigations have determined that 10.4 million people die each year from this condition [1, 2], and the World Health Organization (WHO) estimates that the prevalence of hypertension is 1.13 billion people and that by 2025, it will increase between 15 and 20%, being about 1.5 billion people [3, 4].

In Mexico, this problem is no different, in 2016 the National Survey of Health and Nutrition of Half Way established that one in four adults suffers from hypertension, which corresponds to 25.5% of the population, with approximately 40% being unaware of having said disease, and only 30% of the patients know that they have hypertension, are controlled and the other 30% of the population know that they are sick, but are not controlled [5].

In these times, when the COVID-19 pandemic afflicts us, it is of utmost importance to take care of the health of the population and to have good control of hypertension, because people with this comorbidity are the ones that affect them in a great way sickness, leading even to death [6, 7]. Hybrid intelligent systems have been presented with different applications and shown to be excellent tools for solving complex problems [8–10], since this can use more than two soft computing techniques to solve the same problem, and with this they reduce the computational complexity.

Soft computing is defined as a set of methods that are designed to emulate one or more aspects of biological or social systems, to reproduce knowledge. For this book, artificial neural networks are used, which are inspired by the human biological neuron and fuzzy inference systems, which allow expert systems to reason with uncertainty, as the people do naturally.

The main goal of this research is to create a hybrid neural model to perform the trend analysis of complex systems behavior and thus provide a risk diagnosis, this, using artificial neural network and fuzzy system, in addition to using bio-inspired algorithms to improve model performance.

© The Author(s), under exclusive license to Springer Nature Switzerland AG 2022   1
P. Melin et al., *Nature-inspired Optimization of Type-2 Fuzzy Neural Hybrid Models for Classification in Medical Diagnosis*, SpringerBriefs in Computational Intelligence, https://doi.org/10.1007/978-3-030-82219-4_1

As specific goals, the following are listed:

- Model the behavior trends of complex systems through a neuro-fuzzy model.
- Implement a neural model to learn risk behavior in complex systems.
- Develop Interval Type-2 Fuzzy Systems for classification.
- Carry out a comparison of results of Type-1 and Interval Type-2 fuzzy inference systems.
- Optimize all proposed neuro-fuzzy hybrid models.

The justification for this work is based on the fact that the proposed model can provide an accurate and timely diagnosis of the risk of an extraordinary event occurring based on the information provided, to prevent contingency situations. For this reason, hybrid intelligent systems are applied because they improve their efficiency and reduce the complexity of the problem to be solved in the proposed model.

Soft computing has been previously used for the diagnosis of different illness, which are described as follows.

For the detection of malignant melanoma Warsi et al. [11] propose a new method where dermatoscopic image characteristics are extracted, called multi-direction 3D color texture feature. For this detection, a backpropagation multilayer neural network classifier was used. The proposed method is tested with the PH2 data set, with the neural network classifier accuracy of 97.5% was obtained.

Sejdinović et al. [12] use an artificial neural network for the classification of prediabetes and type 2 diabetes in patients. As input, the information about the Fasting Plasma Glucose (FPG) and a blood test called HbA1c are used, and as output, the classification of the patient health stage is obtained, that is, if it is a healthy patient, if it has prediabetes or Type 2 diabetes. Different tests are performed in which 94.1% accuracy is obtained in the classification of prediabetes, while in the classification of type 2 diabetes an accuracy of 93.3% was obtained.

For the detection of lung diseases, Varela and Melin [13] propose the use of different soft computing techniques. To carry out the study, they use x-ray images that segment and apply a process of extraction of texture characteristics to be able to classify the different diseases. Said classification is carried out using a neural network, with which good results are obtained.

Nour et al. [14] propose an intelligent method based on convolutional neural networks for the detection of COVID-19. For this, x-ray images of the thorax are used and what is sought is to extract distinctive characteristics to determine the disease. With the extracted characteristics, they feed machine learning algorithms such as the nearest k-neighbor, the support vector machine (SVM), and the decision tree. The result is that the SVM was the algorithm that provided the best results with 98.97% precision, 89.39% sensitivity, and 99.75% specificity.

Because the process of diagnosing prostate cancer can be complex, Udoh et. al. [15] propose the use of an Adaptive Neuro-Fuzzy Inference System (ANFIS) which is given different input attributes such as age, pain when urinating, blood in semen, and pelvic pain. As result, they determine that they obtain a 95% correct diagnosis.

For this research, a neuro-fuzzy hybrid system is designed to provide a timely diagnosis of the risk that a patient has in developing different illness such as hypertension and some cardiovascular event based on the behavior of his blood pressure.

This book is organized as follows: In Chap. 2 the theory of soft computing and medical terms are presented, Chap. 3 presents the proposed model to obtain the medical diagnosis, in Chap. 4 presents the study cases to test the different optimization performed to neuro fuzzy hybrid model are presented, and Chap. 5 the conclusions of the proposed model and future work is presented.

References

1. GBD 2017 Risk Factor Collaborators, Global, regional, and national comparative risk assessment of 84 behavioural, environmental and occupational, and metabolic risks or clusters of risks for 195 countries and territories, 1990–2017: a systematic analysis for the Global Burden of Disease Stu. Lancet (London, England). **392**(10159), 1923–1994 (2018)
2. U. Thomas et al., 2020 International Society of Hypertension Global Hypertension Practice Guidelines. Hypertension **75**(6), 1334–1357 (2020)
3. World Health Organization (2017), https://www.who.int/news-room/fact-sheets/detail/cardiovascular-diseases (cvds). Accessed 03 Dec 2020
4. A. Zanchetti et al., 2018 ESC/ESH Guidelines for the management of arterial hypertension. Eur. Heart J. **39**(33), 3021–3104 (2018)
5. I. Campos-Nonato, L. Hernández-Barrera, A. Pedroza-Tobías, C. Medina, S. Barquera, Hipertensión arterial en adultos mexicanos: prevalencia, diagnóstico y tipo de tratamiento. Ensanut MC 2016. Salud Pública de México **60**, pp. 233–243 (2018)
6. E.L. Schiffrin, J.M. Flack, S. Ito, P. Muntner, R.C. Webb, Hypertension and COVID-19. Am. J. Hypertens. **33**(5), 373–374 (2020). https://doi.org/10.1093/ajh/hpaa057
7. COVID-19 and hypertension: What we know and don't know—American College of Cardiology (2020), https://www.acc.org/latest-in-cardiology/articles/2020/07/06/08/15/covid-19-and-hypertension. Accessed 08 Dec 2020
8. M.L. Lagunes, O. Castillo, J. Soria, Methodology for the optimization of a fuzzy controller using a bio-inspired algorithm, in *Fuzzy Logic in Intelligent System Design* (2018), pp. 131–137
9. M. Pulido, P. Melin, G. Prado-Arechiga, A new method based on modular neural network for arterial hypertension diagnosis, in *Nature-Inspired Design of Hybrid Intelligent Systems*, ed. by P. Melin, O. Castillo, J. Kacprzyk (Springer International Publishing, Cham, 2017), pp. 195–205
10. E. Bernal, O. Castillo, J. Soria, F. Valdez, Optimization of fuzzy controller using galactic swarm optimization with type-2 fuzzy dynamic parameter adjustment. Axioms **8**(1) (2019)
11. F. Warsi, R. Khanam, S. Kamya, C.P. Suárez-Araujo, An efficient 3D color-texture feature and neural network technique for melanoma detection. Informatics Med. Unlocked. **17**, 100176 (2019)
12. D. Sejdinović et al., Classification of prediabetes and type 2 diabetes using artificial neural network. CMBEBIH **2017**, 685–689 (2017)
13. S. Varela-Santos, P. Melin, classification of X-Ray images for pneumonia detection using texture features and neural networks, in *Intuitionistic and Type-2 Fuzzy Logic Enhancements in Neural and Optimization Algorithms: Theory and Applications*, ed. by O. Castillo, P. Melin, J. Kacprzyk (Springer International Publishing, Cham, 2020), pp. 237–253

14. M. Nour, Z. Cömert, K. Polat, A novel medical diagnosis model for COVID-19 infection detection based on deep features and Bayesian optimization. Appl. Soft Comput. **97**, 106580 (2020)
15. S.S. Udoh, U.A. Umoh, M.E. Umoh, M.E. Udo, Diagnosis of prostate cancer using soft computing paradigms diagnosis of prostate cancer using soft computing paradigms. Glob. J. Comput. Sci. Technol. D Neural Artif. Intell. **19**(2), 19–26 (2019)

Chapter 2
Theory of Soft Computing and Medical Terms

In this chapter, important concepts to understand better the methods presented in the book are described, such as explaining the different soft computing techniques used, as well as explaining the metaheuristics used to perform the optimization. In addition to defining important medical concepts, such as blood pressure and hypertension, which are concepts that envelop this research.

2.1 Hybrid Systems

Hybrid systems are defined as the combination of two or more soft computing techniques to solve the same problem, in this way reducing its complexity, aiming to improve the efficiency and power of reasoning, as well as the expression of isolated intelligent systems [1]. Initially, the research and development of hybrid systems are focused on combining expert systems and neural networks, with these different useful applications have been developed. Research on the integration of intelligent systems has advanced considerably, and some models and guidelines are beginning to emerge for the development of hybrid neural networks and expert systems applications. Today, fuzzy logic, genetic algorithms, and case-based reasoning have gained attention individually and in combination with other intelligent technologies [2].

Hybrid systems with artificial neural networks and fuzzy systems are divided into the following categories:

- Loose and strong coupling of separate neural and fuzzy modules to carry out specific functions in a complex system.
- Expansion of fuzzy control systems to include neural network components.
- Adjustment of supervised and unsupervised neural networks using fuzzy systems to improve performance.
- Application of fuzzy systems as a modifier of artificial neural networks.

• Using artificial neural networks to improve fuzzy systems [2].

Below is a brief explanation of the mentioned soft computing techniques.

2.1.1 Artificial Neural Networks

An artificial neural network is defined as an information processing system that has performance characteristics in common with biological neurons. These have been developed as generalizations of mathematical models of human cognition or neural biology [3], based on the following assumptions:

1. Information processing occurs in many simple elements called neurons.
2. Signals are transmitted between neurons through connecting links.
3. Each connecting link has an associated weight, which, in a typical neural network, multiplies the transmitted signal.
4. Each neuron has an activation function (usually non-linear) applied to its net input (sum of the weighted input signals) to determine its output [3].

2.1.2 Type-1 Fuzzy Systems

Fuzzy logic is a type of logic that involves approximate modes of reasoning instead of exact ones [4]. It aims to provide grounds for approximate reasoning using imprecise propositions based on fuzzy set theory [5]. Fuzzy logic is a theory that has been implemented in the scientific-technical field and which is extremely useful if we want a certain device (machine, program, application, etc.) to "think" as it would the human mind. Fuzzy logic is fundamentally based on creating a mathematical relationship between an element and a certain fuzzy set for a computer to be able to make an assessment similar to how humans do. Fuzzy Logic born from the publication "Fuzzy Sets" [6] written and proposed by Loftí A. Zadeh for Information and Control journal in the year 1965, based on the work of Lukasiewicz [5] on multivalued logic. Fuzzy logic tries to generate mathematical approximations to solve different types of problems. They claim to produce exact results from imprecise data, which is why it is particularly useful in electronic or computational applications. The fuzzy adjective applied to this logic is because the "non-deterministic" truth values used in it generally have a connotation of uncertainty [4].

2.1.3 Type-2 Fuzzy Logic

Type-2 fuzzy logic is described as an extension of Type-1 fuzzy systems, which was proposed in 1975 by Zadeh. These were designed in mathematical form to represent the uncertainty and vagueness that the linguistic problems bring and are characterized by membership functions, in which the degree of membership for each part of this is a fuzzy set between [0,1], contrasting a Type-1, where it established that a membership grade is a crisp number between [0,1] [7–10].

Interval Type-2 fuzzy sets have two membership functions, a primary one which represents the degree of membership X and the secondary membership function weights of each Type-1 fuzzy sets, which is represented in Eq. 2.1.

$$\tilde{A} = \{ (x, \mu_{\tilde{A}}(x)) | x \in X \} \tag{2.1}$$

Understanding the above, the footprint of uncertainty (FOU) can be defined as the union of all the primary memberships, which has two membership functions, an upper one (UMF) $\overline{\mu}(x)$ and a lower membership function (LMF) $\underline{\mu}(x)$. These represent Type-1 fuzzy sets and have n embedded sets, and in this case, the degree of membership of each element of a Type-2 fuzzy system is an interval [11].

There are different membership functions to represent fuzzy sets, in Eq. 2.2 the Trapezoidal membership functions are represented, and in Eq. 2.3 the Gaussian membership functions are represented.

$$\tilde{\mu}(x) = \left[\underline{\mu}(x), \overline{\mu}(x) \right] = \text{itrapatype2}(x, [a_1, b_1, c_1, d_1, a_2, b_2, c_2, d_2, \alpha])$$

where $a_1 < a_2, b_1 < b_2, c_1 < c_2, d_1 < d_2$

$$\mu_1(x) = \max\left(\min\left(\frac{x - a_1}{b_1 - a_1}, 1, \frac{d_1 - x}{d_1 - c_1} \right), 0 \right)$$

$$\mu_2(x) = \max\left(\min\left(\frac{x - a_2}{b_2 - a_2}, 1, \frac{d_2 - x}{d_2 - c_2} \right), 0 \right)$$

$$\overline{\mu}(x) = \begin{cases} \max(\mu_1(x), \mu_2(x)) & \forall x \notin (b1, c2) \\ 1 & \forall x \in (b1, c2) \end{cases}$$

$$\underline{\mu}(x) = \min(\alpha, \min(\mu_1(x), \mu_2(x))) \tag{2.2}$$

$$\tilde{\mu}(x) = \left[\underline{\mu}(x), \overline{\mu}(x) \right] = \text{igaussatype2}(x, [\sigma, m, \alpha])$$

$$\underline{\mu}(x) = \alpha \exp\left[-\frac{1}{2}\left(\frac{x-m}{\sigma}\right)^2 \right] \text{ Where } 0 < \alpha < 1$$

$$\overline{\mu}(x) = \exp\left[-\frac{1}{2}\left(\frac{x - m}{\sigma}\right)^2 \right] \tag{2.3}$$

2.1.4 *Optimization*

Optimization is an important concept since with this a significant improvement can be made to various applications and is describes as a mathematical process that seeks the best solution to a particular problematic [12]. Optimization has been widely used to improve results in different areas such as control [13], underwater image processing [14], heat production [15], pharmaceutical tableting processes [16], time series prediction [17], among others.

In the case of soft computing techniques, it is common to use bio-inspired algorithms to carry out this process. Furthermore, in recent years, new metaheuristics have been proposed as Emperor penguin optimizer [18], Barnacles Mating Optimizer [19], Sooty Tern Optimization Algorithm [20], Squirrel search algorithm [21], Tunicate Swarm Algorithm [22], to mention a few. The bio-inspired algorithms used in this work are presented in the following sections.

2.1.5 *CSO*

The Chicken Swarm Optimization (CSO) algorithm was proposed in 2014 by Meng [23], which is inspired on trying to imitate the order and behavior in the swarm of chickens and where there will be a division of groups where each one will have a rooster, and some hens and chickens. According to the chicken, it follows different laws of motion, where they will compete with each other according to a specific hierarchical order.

To simplify the actual behavior of chickens, the following rules are proposed:

1. As mentioned, the swarm has different groups, which are made up of a leading rooster, a couple of hens, and chickens.
2. The division of the swarm of chickens will depend on their fitness value, where the chicken with the best fitness will act as the rooster, which will be the head of the group, and the chickens with the worst fitness value will be the chickens. The rest will be hens, which randomly choose which group they correspond to. In addition to the fact that the mother–child relationship is also established in a random form.
3. The hierarchy, dominance, and mother–child relationship will not change, they will only be updated every few steps of time (G).
4. Chickens follow their group-mate rooster in search of food while preventing them from eating their food. Chickens are supposed to randomly steal good food found by others, whereby the chicks will search for food around their mother. Dominant chickens have the upper hand in the competition for food [23].

Chicken movements can be defined mathematically as follows:

Chickens with the best fitness value will have priority in access to food than those with the worst fitness value, that is, the roosters with the best fitness value can perform the search in an ampler space than the roosters with the worst fitness value, which is mathematically translated as follows:

$$x_{i,j}^{t+1} = x_{i,j}^t * \left(1 + Rand\left(0, \sigma^2\right)\right) \tag{2.4}$$

$$\sigma^2 = \begin{cases} 1, & \textit{iff}_i \leq f_k, \\ \exp(\frac{f_k - f_i}{|f_i| + \varepsilon}), & \textit{otherwise}, \end{cases} \quad k \in [1, N], k \neq i \tag{2.5}$$

where:

- $Rand\left(0, \sigma^2\right)$: this is a Gaussian distribution that has a mean of 0 and a standard deviation of σ^2.
- ε: this is used to avoid the error in zero division, which corresponds to a small constant in the computer.
- k: this is an index of roosters, which is chosen randomly by the group of roosters
- f: this is the fitness value of the corresponding x.

Chickens can follow companion roosters for food. In the same way, they can randomly steal the good food found by other chickens, but they can also be suppressed by the other chickens. Dominant hens have the advantage over submissive hens in competing for food. This behavior can be viewed mathematically as follows:

$$x_{i,j}^{t+1} = x_{i,j}^t + S1 * Rand * \left(x_{r1,j}^t - x_{i,j}^t\right) + S2 * Rand * \left(x_{r2,j}^t - x_{i,j}^t\right) \tag{2.6}$$

$$S1 = \exp\left(\frac{f_i - f_{r1}}{abs(f_i) + \varepsilon}\right) \tag{2.7}$$

$$S2 = \exp((f_{r2} - f_i)) \tag{2.8}$$

where:

- $Rand$: are random numbers in the [0,1] interval
- $r1 \in [1, \ldots N]$: is an index of the roosters, which is the group mate of the i-th hen.
- $r2 \in [1, \ldots N]$: is an index of chickens (roosters or hens) which are randomly chosen by the swarm and where $r1 \neq r2$.

Obviously, $f_i > f_{r1}, f_i > f_{r2}$, thus S2 < 1 < S1. Be assume $S1 = 0$ then the i-th hen would search for food followed by other chickens. As the difference in the fitness values of the chickens is greater, $S2$ will be smaller, also the distance between the position of two chickens will be greater. That is, chickens will not easily steal food that has been found by other chickens.

For simplicity, the fitness values of the chickens concerning the fitness value of the rooster are simulated as the competitions between chickens in a group. Suppose

---

Chicken Swarm Optimization

1:   *Initialize a population of N chickens and define the related parameters*
2:   *Evaluate the N chickens' fitness values, t=0;*
3:   **while** *(t < Max_Generation)*
4:     **if** *(t % G== 0)*
5:       *Rank the chickens' fitness values and establish a hierarchal order in the swarm;*
6:       *Divide the swarm into different groups, and determine the relationship between the chicks and*
7:       *mother hens in a group;*
8:     **end if**
9:   **for** *i = 1 : N*
10:     **if** *i == rooster Update its solution/location using equation (1);* **end if**
11:     **if** *i == hen Update its solution/location using equation (3);* **end if**
12:     **if** *i == chick Update its solution/location using equation (6);* **end if**
13:     *Evaluate the new solution;*
14:     *If the new solution is better than its previous one, update it;*
15:   **end for**
17: **end while**

Fig. 2.1 *Chicken Swarm Optimization Pseudocode*

$S2 = 0$, then the *i-th* hen searches for food in its own search space. For the specific group, the fitness value of the rooster is unique. That is, the smaller the fitness value of the *i-th* hen, the $S1$ approaches 1, and the smaller the distance between the positions of the *i-th* hen and her group companion rooster. In such a way that the most dominant hens would have a greater probability of eating the food than the most submissive. The chicks move around their mother in search for food, this movement is translated as follows:

$$x_{i,j}^{t+1} = x_{i,j}^{t} + FL * \left(x_{m,j}^{t} - x_{i,j}^{t} \right) \tag{2.9}$$

where:

- $x_{m,j}^{t}$: Is the position of the *i-th* mother hen $m \in [1, N]$.
- $FL(FL \in (0, 2))$: Determines that the chicken will follow its mother in search of food. Individual differences must be considered, where the FL of each chick is chosen randomly between 0 and 2.

The CSO pseudocode is presented in Fig. 2.1.

2.1.6 CSA

The Crow Search Algorithm (CSA) was proposed by Askarazadeh in 2016 [24], which mimics the behavior of crows. These animals are considered the most intelligent birds since they contain the largest brain with respect to the size of their body. If the brain-body comparison is made, it is slightly lower than that of a human.

Crow Search Algorithm

1:   *Randomly initialize the position of a flock of N crows in the search space*
2:   *Evaluate the position of the crows*
3:   *Initialize the memory of each crow*
4:   **while** *iter* < *iter_max*
5:     **for** i = 1 : N (all N crows of the flock)
6:       *Randomly choose one of the crows to follow (for example j)*
7:       *Define an awareness probability*
8:        *if* $rj \geq AP^{j,iter}$
9:            $x^{i,iter+1} = x^{i,iter} + r_i \times fl^{i,iter} \times (m^{j,iter} - x^{i,iter})$
10:       *else*
11:           $x^{i,iter+1} =$ *a random position of search space*
12:       *endif*
13:     *end for*
14:   *Check the feasibility of new positions*
15:   *Evaluate the new position of the crows*
16:   *Update the memory of crows*
17:   *end while*

Fig. 2.2 Crow search algorithm pseudocode

Crows have been shown to have mirror self-awareness and tool-making skills and can remember faces and warn each other if another unfriendly animal approaches. They can also use tools, communicate in a sophisticated way, and remember where to hide their food for several months. It is well known that crows observe other birds learn the hiding places of their food and then steal it.

If a crow has stolen some food, it will take extra precautions like moving its food to different hiding places to avoid being a victim. They use their own thief experience to predict a thief's behavior and thereby determine the safest way to protect their hiding places. The pseudocode of the algorithm is presented in Fig. 2.2.

The principles of the CSA are:

- Crows live in flocks
- Crows memorize the position of their hiding places
- Crows follow each other to steal
- Crows protect their hiding places from being robbed by a probability

Assume that there is a d-dimensional space, where several crows are included. The number of crows (flock size) is N and the position of the crow i in time (iteration) iter in the search space is defined by a vector $x^{i,iter}$ ($i = 1, 2, \ldots N$; $iter = 1, 2, \ldots, iter$) where $x_1^{i,iter} = \left[x_1^{i,iter}, x_2^{i,iter}, \ldots, x_d^{i,iter} \right]$ and x_{max} is the maximum number of iterations. Each crow memorizes the position of its hiding place. In the iteration, the position of the hiding place of the crow i is obtained at the moment. The crows move within the search space to search for the best food source.

It is assumed that in the iteration *iter*, the crow j wants to visit its hiding place $m^{j,iter}$, in this iteration, crow i decides whether to follow crow j to crow j hiding place. This behavior can take two states:

State 1: Crow j does not know which crow is following him, as a result, crow i approaches hiding place j. The new position of the crow i is obtained by:

$$x^{i,iter+1} = x^{i,iter} + r_i \times fl^{i,iter} \times \left(m^{j,iter} - x^{i,iter}\right) \qquad (2.10)$$

where:

- r_1: is a random number with a uniform distribution between 0 y 1.
- $fl^{i,iter}$: is the length of crow i in iteration *iter*.
- The values of fl mean the local search (in the neighborhood $x^{i,iter}$) and large values are global searches (far from $x^{i,iter}$).

State 2: Crow j knows that crow i is following him, resulting in behavior to protect his hiding place from being robbed. Crow j will trick crow i by changing its position in the search space, both states can be expressed as follows:

$$x^{i,iter+1} = \begin{cases} x^{i,iter} + r_i \times fl^{i,iter} \times \left(m^{j,iter} - x^{i,iter}\right) r_j \geq AP^{j,iter} \\ arandompositionotherwise \end{cases} \qquad (2.11)$$

where:

- r_j: is a random number with a uniform distribution between 0 a 1 and
- $AP^{j,iter}$: is the probability of consciousness of crow j in the iteration *iter*.

In this algorithm, intensification and diversification are mainly controlled by the probability of consciousness (AP) parameter. By decreasing the knowledge probability value, the algorithm searches locally where a current good solution is found in this region. This means that the use of small AP values increases the escalation. Otherwise, as the awareness probability value increases, the probability of searching in the vicinity of good current solutions decreases and the algorithm tends to explore the search space on a global scale. This means that using large AP assures increasing the diversification

2.1.7 FPA

The Flower Pollination Algorithm (FPA) [25] is inspired by the plant pollination process. The objective of this process in biological evolution is the survival of the fittest and the optimal reproduction of the plants, which can be viewed as an optimization process of the plants.

Pollination can be done in two ways:

- Biotic: This is when the pollen is transferred by pollinators, such as insects and animals, and this is carried out in about 90% of the plants.
- Abiotic: This process does not require any pollinator, the wind, and water diffusion help to pollinate, and this is done in about 10% of the plants.

In the same way, pollination can be achieved in the following ways:

- Auto pollination: this refers to the fertilization of a flower, using the pollen of the same flower or different flowers of the same plant, and this occurs when there is no reliable pollinator available, an example of this is the blossoming peach.
- Cross-pollination: This can occur from the pollen of a flower of a different plant.

Other adjustments that the algorithm can have: to simplify the pollination process, the author assumes that each plant only has one flower, and each flower only produces one pollen gamete, which is why it is not necessary to distinguish it from a gamete, a plant, a flower or the solution of a problem [25].

1. Once knowing the characteristics of this process, the following rules are created to be expressed in the algorithm: Biotic and cross-pollination are considered as the global process of pollination, with pollinators that transport pollen through Lévy flights.
2. Abiotic pollination and self-pollination are considered local pollination.
3. The constancy of the flowers can be considered as the probability of reproduction, which is proportional to the similarity of the two flowers being involved.
4. Local and global pollination are controlled by a probability of change $p \in [0, 1]$. Due to physical proximity and other factors, such as wind, local pollination can have a significant fraction in p in all pollination activities. In Fig. 2.3, the pseudocode of the FPA is presented.

As mentioned above, global pollination is the one that occurs through pollinators, such as insects, and where pollen can travel long distances, because insects can travel in this way. This ensures that the pollination and reproduction of the fittest, and is represented by g. The mathematical representation of the described rules is presented below.

The first rule, plus the constancy of the flower is represented as:

$$x_i^{t+1} = x_i^t + L(x_i^t) - g, \tag{2.12}$$

where x_i^t is the pollen i or the solution of the vector x_i in the iteration t, and g_* is the best current solution found among all the solutions in the current iteration. The parameter L is the force of pollination; this means that it is the strength of the steps for random flight purposes. Because insects can move over long distances, Lévy's flight can be used to efficiently mimic that characteristic, that is, $L > 0$ is drawn for a Lévy distribution.

Flower Pollination Algorithm

1:   *Objective* min *or* max *f*(x), x=(x_1, x_2, ..., x_d)
2:   *Initialize a population of n flowers/pollen gametes with random solutions*
3:   *Find the best solution* g_* *in the initial population*
4:   *Define a switch probability* $p \in [0, 1]$
5:   **while** ($t <$ *MaxGeneration*)
6:       **for** $i = 1 : n$ *(all n flowers in the population)*
7:           **if** *rand* $< p$,
8:               *Draw a (d-dimensional step vector L which obeys a Lévy distribution)*
9:               *Global pollination via* $x_i^{t+1} = x_i^t + L(g_* - x_i^t)$
10:         **else**
11:             *Draw* ϵ *from a uniform distribution in [0, 1]*
12:             *Randomly choose j and k among all the solutions*
13:             *Do local pollination via* $x_i^{t+1} = x_i^t + \epsilon(x_j^t - x_k^t)$
14:         **end if**
15:         *Evaluate new solutions*
16:         *If new solutions are better, update them in the population*
17:     **end for**
18:         *Find the current best solution* g_*
19:   **end while**

Fig. 2.3 Flower pollination algorithm pseudocode

$$L \sim \frac{\lambda \Gamma(\lambda) \sin(\pi \lambda/2)}{\pi} \frac{1}{s^{1+\lambda}}, \, (s \gg s_0 > 0) \tag{2.13}$$

In Eq. 2.2, $\Gamma(\lambda)$ is the standard gamma function, and this distribution is valid for long steps $s > 0$.

Local pollination and constancy of flowers (Rule 2) are represented as:

$$x_i^{t+1} = x_i^t + \varepsilon\left(x_j^t - x_k^t\right) \tag{2.14}$$

where x_j^t and x_k^t are pollen from different flowers of the same plant. This essentially mimics the constancy of the flower in a limited neighborhood. Mathematically, if x_j^t and x_k^t come from the same species or selected by the same population, this becomes a random local step if ε is drawn for a random distribution [0, 1].

Most flower pollination activities can be done both locally and globally. In practice, patches of adjacent flowers or flowers in the not-so-far neighborhood are more likely to be pollinated by pollen from local flowers than by pollen that is farther away. To do this, we use a probability of change (Rule 4) or a proximity p to change between local and global pollination, in other words, p controls the exploration and the exploitation of the algorithm.

2.1.8 BSA

The Bird Swarm Algorithm (BSA) was proposed in 2015 by Meng [26] and it is based on the behavior of birds in the swarm. These behaviors are social behavior and social interaction and mimic feeding, flying, and vigilance to resolve problems through optimization.

In this algorithm, the social behaviors of birds are summarized by the rules explained as follows:

1. The behaviors of vigilance and foraging can switch in each bird. These behaviors are modeled as a stochastic decision.
2. While foraging, birds can remember and update their and the swarms' best previous experience regarding a patch of food. Remembered experience can be utilized to search for food sources. Regarding social information, this is instantly shared with the entire swarm.
3. While the birds keep vigilance, each one intends to go to the center of the swarm, this behavior can be affected by the interference induced by the competition in all the swarm. When the birds have the higher provisions would be more possible to be closer to the center of the swarm than birds with lower reserves of food.
4. Birds can usually fly to other sites when this is done, birds may change between producing and scrounging, where those birds with the highest reserve of food are taken as a producer, and the one that has the lowest provisions are taken as a scrounger. Birds that take a middle provision are select in the random form to be a producer and scrounger.
5. Producers actively search for food. Scroungers would randomly follow a producer to search for food.

The rules can be expressed in mathematical form as:

(a) Foraging behavior

Each bird searches for food remembering its experience and the experience of swarms, this part is representing as:

$$x_{i,j}^{t+1} = x_{i,j}^{t} + \left(p_{i,j} - x_{i,j}^{t}\right) \times CX \times rand(0, 1) + \left(g_j - x_{i,j}^{t}\right) \times S \times rand(0, 1),$$

$$(2.15)$$

where:

- $j \in [1, \ldots, D]$, $rand(0, 1)$ is independent number distributed in $(0,1)$ uniformly.
- C and S are known as cognitive and social acceleration coefficients respectively and represents two positive numbers.
- $P_{i,j}$ is the best previous position in the i_{th} bird and g_j is the best previous position shared in the swarm.

(b) Vigilance behavior

Each bird would try to change to the center of the swarm and compete with others, for this, each bird would not change in a direct way to the center of the swarm, this behavior is represented mathematically as:

$$x_{i,j}^{t+1} = x_{i,j}^t + A1\left(mean_j - x_{i,j}^t\right) \times rand(0, 1) + A2\left(p_{k,j} - x_{i,j}^t\right) \times rand(-1, 1)$$
(2.16)

$$A1 = a1 \times \exp\left(-\frac{pFit_i}{sumFit + \varepsilon} \times N\right)$$
(2.17)

$$A2 = a2 \times exp\left(\left(\frac{pFit_i - pFit_k}{|pFit_k - pFit_i| + \varepsilon}\right)\frac{N \times pFit_k}{sumFit + \varepsilon}\right)$$
(2.18)

where k is a positive integer, selected in random form between 1 and N. $pFit_i$ denotes the best fitness value in the ith position and $sumFit$ is the sum of the best fitness value of the swarms. ε is a small number used to avoid the zero-division error. $mean_j$ is the jth element of the average place of the whole swarm. a_1 and a_2 are positive constants in [0,2].

(a)   Flight Behavior

Birds may move to other sites in response to foraging, predation threats, or another reason. Once are placed in the new place; the birds search for food again. The producers observe for patches of food, while the scroungers try to feed on the food patch found by the producers. These behaviors are illustrating as follow:

$$x_{i,j}^{t+1} = x_{i,j}^t + \text{rand}n(0, 1) \times x_{i,j}^t,$$
(2.19)

$$x_{i,j}^{t+1} = x_{i,j}^t + \left(x_{k,j}^t - x_{i,j}^t\right) \times FL \times \text{rand}(0, 1),$$
(2.20)

where randn (0, 1) represents a Gaussian distributed random numbers with mean 0 and standard deviation 1, k $\in$ [1,2,3... N], k $\neq$ i. FL (FL $\in$ [0, 2]) represents that the scrounger will follow the producer to search for food.

For simplicity, the author assumes that each bird may fly to another place every FQ unit interval, where FQ is a positive integer. The pseudocode of BSA is presented in Fig. 2.4.

2.2   Blood Pressure

Blood pressure (BP) is defined as the force exerted against the walls of the arteries as the heart pumps blood, this process is necessary and important since it is responsible for the blood to circulate through the blood vessels and supply oxygen and nutrients

Bird Swarm Algorithm

1:   *Input N: the number of individuals (birds) contained by the population*
2:         *M: the maximum number of iterations*
3:         *FQ: the frequency of birds' flight behaviors*
4:         *P: the probability of foraging for food*
5:         *C, S, a1, a2, FL: five constant parameters*
6:   *t=0; Initialize the population and define the related parameters*
7:   *Evaluate the N individuals' fitness value, and find the best solution*
8:   ***While** (t < M)*
9:       ***If** (t % FQ ≠ 0)*
10:        ***For** i = 1 : N*
11:          ***If** rand (0,1) < P*
12:              *Birds forage for food (eq. X)*
13:          ***Else***
14:              *Birds keep vigilance (eq. X)*
15:          ***End if***
16:        ***End for***
17:      ***Else***
18:          *Divide the swarm into two parts: producers and scroungers.*
19:          ***For** i = 1 : N*
20:            ***If** i is a producer*
21:                *Producing (eq. X)*
22:            ***Else***
23:                *Scrounging (eq. X)*
24:          ***End if   End For***
25:      ***End If** Evaluate new solutions*
26:      *if the new solutions are better than their previous ones, update then*
27:        *Find the best solutions*
28:   *t=t+1; **End while***
29:   ***Output**: the individual with the best objective function value in the*

Fig. 2.4 Bird swarm algorithm pseudocode

necessary to all organs so that the body can function in a good way [27, 28]. In clinical form, blood pressure levels are expressed in millimeters of mercury (mmHg), additionally are divided into two components:

Systolic Pressure: is the maximum value, it measures the force of the blood in the arteries when the heart contracts (beats).

Diastolic Pressure: corresponds to the minimum value, and measures the force of the blood in the arteries while the heart is relaxed (filling with blood in between beats) [29, 30].

The arterial system is made up of the large arteries or capacity arteries, which, in tune with the heart rate, distend with each surge of blood, cushioning its pressure; then it contracts in each cycle, propelling the blood towards more peripheral territories,

Table 2.1 Blood pressure classification

Category	Systolic		Diastolic
Optimal	<120	and	<80
Normal	120–129	and/or	80–84
Normal High	130–139	and/or	85–89
Hypertension grade 1	140–159	and/or	90–99
Hypertension grade 2	160–179	and/or	100–109
Hypertension grade 3	≥180	and/or	≥110
Isolated systolic hypertension	≥140	and	<90

such as the small resistance arteries or arterioles, tiny microscopic vessels that do offer great resistance to the passage of blood. The blood reaches them with the pressure already very cushioned and from there the passage of oxygen and nutrients to the tissues is verified [31].

Based on the European guideline for the management of hypertension [32] in adults, normal blood pressure is defined by the following values: a systolic pressure below 139 mmHg and a diastolic pressure below 89 mmHg. The different values of blood pressure are presented in Table 2.1 [32]. It is normal for blood pressure to change, being lower at night with sleep and higher in the early hours of the morning. In activity, the blood pressure usually goes up. However, once the activity is stop, the blood pressure goes back to the normal range. Generally, blood pressure increases based on age and body size. Newborn babies often have very low blood pressure values that are considered normal for them. On the other hand, adolescents have values similar to that of adults [33].

2.2.1 Hypertension

Hypertension is defined as the maintained elevation of blood pressure above the limits considered normal [34–36]. The disease becomes fatal due to the damage it produces in the impact organs, which are:

- Blood vessels: arteriosclerosis and atherosclerosis.
- Heart: heart failure.
- Kidney: kidney failure.
- Eyes: blindness.

Hypertension is classified as:

Essential hypertension: develops without a specific cause, but with a hereditary history, appears in isolation, or is part of a complex set of alterations found around insulin resistance [37].

Secondary hypertension: In this type of hypertension, there are demonstrable causes, where the most frequent are vascular and endocrine diseases, pregnancy, acute stress, and aortic coarctation. It should be suspected in two situations:

- Hypertension in people with 35 years or less
- Absence of a family history of high blood pressure [38, 39].

In some cases, the elevation of blood pressure is reversible when the disease is successfully treated.

The risk factors related to the development of hypertension are:

1. Excessive alcohol consumption or tobacco use.
2. Overweight and physical inactivity.
3. Age and gender, 35-year-old men are more likely to develop hypertension than women, but when women enter menopause, this probability is similar.
4. High salt intake.
5. Sleep apnea.
6. Family history, if your parents or other close relatives have this disease, it is likely that you will develop hypertension.
7. Race, African Americans have more probability to develop hypertension [33].

2.2.2 Heart Rate

Heart Rate is understood as the number of the heart contractions in a time, which corresponds to the number of beats in one minute. Age can produce changes in the regularity and the velocity of the heart rate, this can be translated into a heart or other condition which should be treated by a specialist [40].

When people are at rest, the heart pumps the least amount of blood necessary, this being between 60 and 100 beats per minute (BPM) in a healthy person, but it should be mentioned that a heart rate less than 60 BPM does not mean that the patient has some medical problem as it can be due to some medication or a well-trained athlete, since, in them, the heartbeat is usually up to 40 beats per minute. Heart rate values according to age and gender are presented in Table 2.2 [41, 42].

Two conditions depend on the heart rate:

Bradycardia: When the heart rate is less than 60 beats per minute, this is when the heart rate is slow or irregular, which may cause dizziness or difficulty breathing in activities in normal form or light exercises [43].

Tachycardia: When the heart rate is greater than 100 beats per minute, that is the heart rate is fast or irregular. When this condition happens, the heart cannot effectively pump blood with high levels of oxygen to the body [44].

Table 2.2 Values of heart rate according to age and gender

	Men				Women			
Age	Ill	Normal	Good	Excellent	Ill	Normal	Good	Excellent
20–29	86 or more	70–84	62–68	60 or less	96 or more	78–94	72–76	70 or less
30–39	86 or more	72–84	64–70	62 or less	98 or more	80–96	72–78	70 or less
40–49	90 or more	74–88	66–72	64 or less	100 or more	80–98	74–78	72 or less
50 or more	90 or more	76–88	68–74	66 o or less	104 or more	84–102	76–86	74 or less

2.2.3   Nocturnal Blood Pressure Profile

With the study carried out with the Ambulatory Blood Pressure Monitoring, the variability of blood pressure over 24 h is observed, obviously including the behavior of blood pressure during the rest period and thus obtain a more precise diagnosis [45].

The normal circadian profile is characterized by a 10–20% decrease in nocturnal BP figures versus daytime or activity BP figures. The absence of a decrease in night-time BP figures <10% is considered a *non-dipper* pattern. Another way to define the *dipper / no dipper* pattern is through the night/day ratio so that the *dipper* patients would present a ratio between 0.90 and 0.80, the *non-dippers* between 0.91–1.00, the *dipper extreme* (nocturnal BP decrease >20% of the daytime BP figures) is <0.80 and the *riser* (mean of the nocturnal BP values higher than the daytime BP) has a ratio >1.00 [46].

The absence of a nocturnal decrease in BP, the *non-dipper* pattern, has classically been associated with a higher risk and worse cardiovascular prognosis than the *dipper* pattern. Recent studies reaffirm the low reproducibility of the *dipper / non-dipper* pattern. A review by Mc Gowan et al. concluded that, in the long term, the nocturnal decrease in BP expressed as a continuous variable was a much more stable parameter and a better predictor of cardiovascular risk than the *dipper / no dipper* pattern [47, 48].

2.2.4   Ambulatory Blood Pressure Monitoring

Ambulatory Blood Pressure Monitoring (ABPM) is a non-invasive study to obtain blood pressure measurements during 24 h, which is described as a device connected to a blood pressure cuff that keeps records of arterial pressure and pulse of patients in a defined time interval, which ranges from 15 to 20 min in the day and 30 min at night, to be later transferred to a computer. Hypertension is determined when more than 40% of the records are greater than 135/85 mmHg during the day and greater than 120/70 mmHg at night [45, 49].

Medical cardiologists recommend using ABPM in the following cases:

- Suspicion of white coat hypertension: This is when the blood pressure is high in the office, but at home these figures are normal, this is because many patients tend to get nervous thinking that the doctor will give them the diagnosis of some serious illness [50].
- Normal blood pressure figures during the medical visit, but with signs of impact organ damage.
- Sporadic or crisis hypertension.
- • Evaluation of the blood pressure profile (dipper or non-dipper).
- Hidden or masked hypertension, patients with normal blood pressure figures in the office while the mean home ABPM values are in the hypertensive range [26].

2.2.5 Framingham Heart Study

The Framingham Heart Study started in 1948 [51], directed by the National Cardiac Institute, which is carried out to identify risk factors that have been the cause for the development of cardiovascular diseases.

This study commenced by enlisting a group of 5,209 men and women from Framingham, aged between 30 and 62 years old, who did not have any symptoms of cardiovascular disease or suffered a heart attack or stroke. During the study, different groups of people have been added, such as the offspring of the first group in 1971, the multicultural Omni group in 1994, the third generation of the offspring's in 2002, and the second group of Omni in 2003 [51].

Different risk functions have been developed to identify other diseases, as arterial fibrillation, coronary disease, diabetes, among others.

One of the diseases that is included in these risk functions is hypertension, with this, which is a Weibull regression model [52, 53], it can find the percentage of risk that a patient has in developing this condition in 4 years. To obtain the diagnosis, patient information is used, such as Age, Systolic Blood Pressure (SBP), Diastolic Blood Pressure (DBP), Body Mass Index (BMI), Sex, Smoking, Parental Hypertension. The mathematical expression for the risk of developing hypertension is the following:

$$FHSpredictorrisk = 1 - exp\left[-exp\left(\frac{\ln(4) - 22.94954 + \sum_{i=1}^{8} x_i}{0.8769}\right)\right] \quad (2.21)$$

It can be observed that different parameters are used:

- ln (4) determines the risk time, in this case, 4 years.
- 22.94954 is a constant that corresponds to the intercept.
- 0.8769 is a constant that corresponds to the scale.

- In addition, each risk variable is assigned a β regression coefficient and x is the level of each variable and is assigned as follows:
- $x_1 = -0.15641*Age$, $x_2 = -0.20293*Sex$, $x_3 = -0.05933*SBP$ $x_4 = -0.12847*DBP$, $x_5 = -0.19073*Smoking$, $x_6 = -0.16612*Parental$ $Hypertension$, $x_7 = -0.03388*BMI$, $x_8 = 0.00162*DBP*Age$.
- If the patient smokes, then the variable Smoking is assigned to one, otherwise it is assigned to 0.
- If none of the parents are hypertensive the variable Parental Hypertension is assigned to 0, otherwise if one of the parents is hypertensive it is assigned to 1 and finally, if both of the parents are hypertensive it is assigned to 2 [51, 54].

2.2.6 Cardiovascular Risk

Cardiovascular risk is the probability that people can have a cardiovascular event in a time, which can be between 5 and 10 years [34]. The main risk factors can be:

Non-modifiable: age, sex, genetic factors, family history.

Modifiable: high blood pressure, smoking, hypercholesterolemia, diabetes mellitus, obesity, physical inactivity.

The last ones are those of greatest interest to medical cardiologists since by acting preventively they can be reduced [55].

References

1. P. Melin, O. Castillo, *Hybrid Intelligent Systems for Pattern Recognition Using Soft Computing*, 1st edn. (Springer-Verlag, Berlin, Heidelberg, 2005)
2. L.R. Medsker, Overview of intelligent systems, in *Hybrid Intelligent Systems*, ed. by L.R. Medsker (Springer US, Boston, MA , 1995), pp. 1–15
3. D. Graupe, *Principles of Artificial Neural Networks: Basic Designs To Deep Learning (4th Edition)* (World Scientific Publishing Company, Chicago, 2019)
4. O. Castillo, P. Melin, Type-1 fuzzy logic, in *Type-2 Fuzzy Logic: Theory and Applications*, ed. by O. Castillo, P. Melin (Springer Berlin Heidelberg, Berlin, Heidelberg, 2008), pp. 5–28
5. G. Chen, T.T. Pham, N. Boustany, *Introduction to Fuzzy Sets, Fuzzy Logic, and Fuzzy Control Systems* (2000)
6. L.A. Zadeh, Fuzzy sets. Inf. Control **8**(3), 338–353 (1965)
7. O. Castillo, P. Melin, Type-2 fuzzy logic, in *Type-2 Fuzzy Logic: Theory and Applications*, ed. by O. Castillo, P. Melin (Springer Berlin Heidelberg, Berlin, Heidelberg, 2008), pp. 29–43
8. O. Castillo, L.T. Aguilar, Background on type-1 and type-2 fuzzy logic, in *Type-2 Fuzzy Logic in Control of Nonsmooth Systems: Theoretical Concepts and Applications*, ed. by O. Castillo, L.T. Aguilar (Springer International Publishing, Cham, 2019), pp. 5–19
9. J.M. Mendel, L. Fellow, General type-2 fuzzy logic systems made simple: a tutorial. IEEE Trans. Fuzzy Syst. **22**(5), 1162–1182 (2014)
10. J.R. Castro, O. Castillo, P. Melin, *An Interval Type-2 Fuzzy Logic Toolbox for Control Applications* (2007)

11. F. Valdez, C. Peraza, O. Castillo, Study cases to test fuzzy harmony search, in *General Type-2 Fuzzy Logic In Dynamic Parameter Adaptation For The Harmony Search Algorithm*, ed. by F. Valdez, C. Peraza, O. Castillo (Springer International Publishing, Cham, 2020), pp. 13–67

12. P. Jain, P. Kar, Non-convex optimization for machine learning. Found. Trends Mach. Learn. **10**, 142–336 (2017)

13. O.R. Carvajal, O. Castillo, J.J. Soria, Optimization of membership function parameters for fuzzy controllers of an autonomous mobile robot using the flower pollination algorithm. J. Autom. Mob. Robot. Intell. Syst. **12**(1), 44–49 (2018)

14. C. Sánchez-Ferreira, L.S. Coelho, H.V.H. Ayala, M.C.Q. Farias, C.H. Llanos, Bio-inspired optimization algorithms for real underwater image restoration. Signal Process. Image Commun. **77**, 49–65 (2019)

15. M. Woźniak, K. Książek, J. Marciniec, D. Połap, Heat production optimization using bio-inspired algorithms. Eng. Appl. Artif. Intell. **76**, 185–201 (2018)

16. H.M. Zawbaa, S. Schiano, L. Perez-Gandarillas, C. Grosan, A. Michrafy, C.-Y. Wu, Computational intelligence modelling of pharmaceutical tabletting processes using bio-inspired optimization algorithms. Adv. Powder Technol. **29**(12), 2966–2977 (2018)

17. D. Sánchez, P. Melin, O. Castillo, Fuzzy dynamic parameter adaptation for particle swarm optimization of modular granular neural networks applied to time series prediction, in *Recent Advances of Hybrid Intelligent Systems Based on Soft Computing*, ed. by P. Melin, O. Castillo, J. Kacprzyk (Springer International Publishing, Cham, 2021), pp. 189–204

18. G. Dhiman, V. Kumar, Emperor penguin optimizer: A bio-inspired algorithm for engineering problems. Knowledge-Based Syst. **159**, 20–50 (2018)

19. M.H. Sulaiman, Z. Mustaffa, M.M. Saari, H. Daniyal, Barnacles mating optimizer: a new bio-inspired algorithm for solving engineering optimization problems. Eng. Appl. Artif. Intell. **87**, 103330 (2020)

20. G. Dhiman, A. Kaur, STOA: a bio-inspired based optimization algorithm for industrial engineering problems. Eng. Appl. Artif. Intell. **82**, 148–174 (2019)

21. M. Jain, V. Singh, A. Rani, A novel nature-inspired algorithm for optimization: squirrel search algorithm. Swarm Evol. Comput. **44**, 148–175 (2019)

22. S. Kaur, L.K. Awasthi, A.L. Sangal, G. Dhiman, Tunicate swarm algorithm: a new bio-inspired based metaheuristic paradigm for global optimization. Eng. Appl. Artif. Intell. **90**, 103541 (2020)

23. X. Meng, Y. Liu, X. Gao, H. Zhang, A new bio-inspired algorithm: chicken swarm optimization, in *Advances in Swarm Intelligence* (2014), pp. 86–94

24. Askarzadeh, A novel metaheuristic method for solving constrained engineering optimization problems: crow search algorithm. Comput. Struct. **169**(Supplement C), 1–12 (2016)

25. X.S. Yang, M. Karamanoglu, X. He, Flower pollination algorithm: a novel approach for multiobjective optimization. Eng. Optim. **46**(9), 1222–1237 (2014)

26. X.-B. Meng, X.Z. Gao, L. Lu, Y. Liu, H. Zhang, A new bio-inspired optimisation algorithm: bird swarm algorithm. J. Exp. Theor. Artif. Intell. **28**(4), 673–687 (2016)

27. American Heart Association (2015), http://www.heart.org/HEARTORG/Conditions/HighBloodPressure/High-Blood-Pressure-or-Hypertension_UCM_002020_SubHomePage.jsp. Accessed 15 Oct 2018

28. M. Paul et al., Measurement of blood pressure in humans: a scientific statement from the American heart association. Hypertension **73**(5), e35–e66 (2019)

29. L.R. Krakoff, Introduction: definition and classification of arterial pressure phenotypes, in *Disorders of Blood Pressure Regulation: Phenotypes, Mechanisms, Therapeutic Options*, ed. by A.E. Berbari, G. Mancia (Springer International Publishing, Cham, 2018), pp. 3–9

30. J. Redon, G. Pichler, F. Martinez, Blood pressure control in europe and elsewhere, in *Manual of Hypertension of the European Society of Hypertension* (CRC Press, 2019), pp. 25–30

31. E. I. Cabrera Fischer, "Structural basis of the circulatory system," Biomechanical Modeling of the Cardiovascular System. IOP Publishing, pp. 1–18, 2019.

32. Zanchetti et al., 2018 ESC/ESH Guidelines for the management of arterial hypertension. Eur. Heart J. **39**(33), 3021–3104 (2018)

33. S.S. Franklin, V. Bell, G.F. Mitchell, Diagnostic and prognostic significance of blood pressure indices, in *Disorders of Blood Pressure Regulation: Phenotypes, Mechanisms, Therapeutic Options*, ed. by A.E. Berbari, G. Mancia (Springer International Publishing, Cham , 2018), pp. 11–21
34. G.L. Bakris, M. Sorrentino, *Braunwald's Heart Disease Family of Books* (Elsevier, 2018), pp. 15–18
35. V. Papademetriou, E.A. Andreadis, C. Geladari, *Management of Hypertension* (Springer International Publishing AG, Cham, 2019)
36. C. Rosendorff, *Essential Cardiology*, 3rd edn. (Springer, Bronx, NY, USA, 2013)
37. R. Bunag, Essential hypertension, in *xPharm: The Comprehensive Pharmacology Reference* (Elsevier Inc., 2007), pp. 1–6
38. N. Li, M. Wang, M. Cao, Summary of secondary hypertension, in *Secondary Hypertension: Screening, Diagnosis and Treatment*, ed. by N. Li (Springer Singapore, Singapore, 2020), pp. 3–21
39. E. Berbari, N.A. Daouk, A.R. Jurjus, Secondary hypertension: infrequently considered aspects—illicit/recreational substances, herbal remedies, and drug-associated hypertension, in *Disorders of Blood Pressure Regulation: Phenotypes, Mechanisms, Therapeutic Options*, ed. by A.E. Berbari, G. Mancia (Springer International Publishing, Cham, 2018), pp. 723–759
40. D.A.S. Silva, T.R. de Lima, M.S. Tremblay, Association between resting heart rate and health-related physical fitness in Brazilian adolescents. Biomed Res. Int. **2018**, 3812197 (2018)
41. Ricarte, Heart rate and blood pressure responses to a competitive role-playing game. Aggress. Behav. **27**(5), 351–359 (2001). http://apps.isiknowledge.com.proxy.lib.sfu.ca/full_record.do?product=WOS&search_mode=GeneralSearch&qid=1&SID=4Cjkj9booJfE2jf618g&page=1&doc=1
42. P. Palatini, Heart rate as a cardiovascular risk factor in hypertension, in *Manual of Hypertension of the European Society of Hypertension* (CRC Press, 2019), pp. 121–126
43. Bradycardia - Harvard Health (2019), https://www.health.harvard.edu/a_to_z/bradycardia-a-to-z. Accessed 08 Dec 2020
44. Tachycardia - Harvard Health (2020), https://www.health.harvard.edu/a_to_z/tachycardia-a-to-z. Accessed 08 Dec 2020
45. A. Dadlani, K. Madan, J.P.S. Sawhney, Ambulatory blood pressure monitoring in clinical practice. Indian Heart J. **71**(1), 91–97 (2019)
46. M.D. Feria-carot, J. Sobrino, Nocturnal hypertension. Hipertens. y riesgo Cardiovasc. **28**(4), 143–148 (2011)
47. M. Brian, A. Dalpiaz, E. Matthews, S. Lennon-Edwards, D. Edwards, W. Farquhar, Dietary sodium and nocturnal blood pressure dipping in normotensive men and women. J. Hum. Hypertens. Hypertens. **31**, 145–150 (2016)
48. S. J. Crinion et al., Nondipping nocturnal blood pressure predicts sleep apnea in patients with hypertension. J. Clin. Sleep Med. **15**(07), 957–963 (2020)
49. K. Kario et al., Diagnostic value of home blood pressure, in *Home Blood Pressure Monitoring*, ed. by G.S. Stergiou, G. Parati, G. Mancia (Springer International Publishing, Cham, 2020), pp. 45–54
50. C. Cuspidi, C. Sala, M. Tadic, G. Grassi, White coat and masked hypertension, in *Disorders of Blood Pressure Regulation: Phenotypes, Mechanisms, Therapeutic Options*, ed. by A.E. Berbari, G. Mancia (Springer International Publishing, Cham, 2018), pp. 599–612
51. Framingham Heart Study (2019), https://www.framinghamheartstudy.org/risk-functions/hypertension/index.php. Accessed 03 Dec 2020
52. C.-D. Lai, Weibull distribution, in *Generalized Weibull Distributions*, ed. by C.-D. Lai (Springer Berlin Heidelberg, Berlin, Heidelberg, 2014), pp. 1–21
53. H. Rinne, *The Weibull Distribution A Handbook* (Taylor & Francis, Boca Raton, FL, 2009)
54. N.I. Parikh et al., A risk score for predicting near-term incidence of hypertension: the Framingham heart study. Ann. Intern. Med. **148**(2), 102–110 (2008)
55. J.M. Lobos Bejarano, C. Brotons Cuixart, Factores de riesgo cardiovascular y atención primaria: evaluación e intervención. Aten. Primaria **43**(12), 668–677 (2011)

Chapter 3
Proposed Model to Obtain the Medical Diagnosis

In this chapter, it is explained how the neuro-fuzzy hybrid model was created to obtain the risk diagnosis of developing hypertension, which describes in detail how each of the soft computing techniques was used, and also how the used database was generated. A description of the created Interval Type-2 fuzzy systems (IT2FS) is presented and then a comparison of the results with the Type-1 fuzzy systems was done to evaluate their performance.

3.1 Neuro-Fuzzy Hybrid Model

A neuro-fuzzy hybrid model is proposed to provide a medical diagnosis based on the blood pressure of patients, the scheme of which is presented in Fig. 3.1, and each of its parts is described.

First, a database is created with the blood pressure records of a group of 300 people, of which 24 belong to professors and graduate students in computer science at the Tijuana Institute of Technology and 276 are patients of the Cardio Diagnostic Center, both located in the city of Tijuana. Ambulatory blood pressure monitoring is used to obtain the record of blood pressure readings, operating two models: Microlife Watch BP03 ABPM monitor and the Spacelab 90217A ABPM monitor. The records are organized into systolic pressure, diastolic pressure, and heart rate. They are the inputs to the modules used in the modular neural network respectively, where 47 readings are taken from each patient for each module to train the information, learn the behavior, the variation from person to person, and thereby obtain the trend of the behavior of blood pressure in 24 h.

© The Author(s), under exclusive license to Springer Nature Switzerland AG 2022 25
P. Melin et al., *Nature-inspired Optimization of Type-2 Fuzzy Neural Hybrid Models for Classification in Medical Diagnosis*, SpringerBriefs in Computational Intelligence, https://doi.org/10.1007/978-3-030-82219-4_3

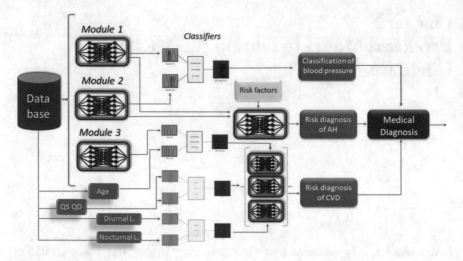

Fig. 3.1 Hybrid neuro-fuzzy model

Different fuzzy systems have a different function of classifying the information, which are described below:

The first fuzzy classifier provides the blood pressure level of the patients, for this, the result of the blood pressure trend obtained by the first and second modules of the modular neural network is given as input. For the second fuzzy classifier, the output of the third module of the modular neural network is used, which corresponds to the heart rate, and together with the age, it gives us the level of the heart rate that the patient has. To obtain the classification of the nocturnal blood pressure profile, the day and night readings are divided, the quotient of systolic and diastolic pressure is obtained so that they are the input to the fuzzy classifier and thus obtain the mentioned result. For the last fuzzy classifier, the percentages of the daytime and nighttime pressure load will be used to obtain the level of the patient's pressure load and with this to be able to determine if is prone to developing a damage in any of the organs.

The next part in the diagram corresponds to a monolithic neural network, which is given 7 risk factors as inputs for a group of people, which are: age, sex, body mass index, systolic and diastolic pressure, if the patient smokes, and if either of the parents is hypertensive. With this information, the artificial neural network is trained, and learns the different risk factors to provide a risk diagnosis of developing hypertension in 4 years.

Next, it has another modular neural network, where each module is given risk factors as input, which are: age, sex, body mass index, systolic pressure, if the patient is undergoing hypertension treatment, if the patient smokes and if it is a diabetic patient. The first module provides the risk of developing a cardiovascular event in 10 years, while modules 2 and 3 provide us with the age of the patient's heart that is with and without treatment, respectively.

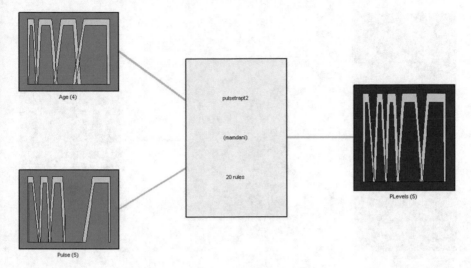

Fig. 3.2 IT2FS for heart rate classification

Taking the results of each of the used intelligent computing techniques, a final medical diagnosis is made. It is worth mentioning that each of the elements of the proposed model was optimized with bio-inspired algorithms to provide an accurate diagnosis, which will be described in the next chapter.

3.2 IT2FS for Classification of Heart Rate Level

An Interval Type-2 fuzzy system is designed for the heart rate classification, to observe which one provides better results. Once the Type-1 fuzzy system has been optimized, the parameters of the best obtained fuzzy system are taken and set the IT2FS to start from there and add the footprint of uncertainty. The general scheme is presented in Fig. 3.2.

3.3 IT2FS for Classification of Nocturnal Bloor Pressure Profile

In the same way as the fuzzy system for heart rate classification, an Interval Type 2 fuzzy system is designed for the classification of the nocturnal blood pressure profile, also taking the best parameters obtained from its optimization. The general scheme is illustrated in Fig. 3.3.

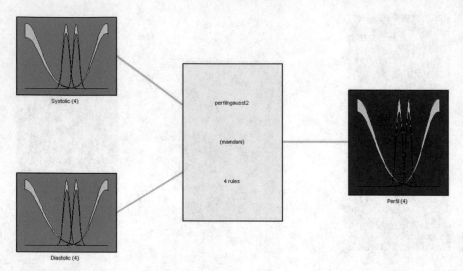

Fig. 3.3 IT2FS for nocturnal blood pressure profile classification

To be able to provide the patient's nocturnal blood pressure profile classification, the Interval Type 2 fuzzy system requires as input the quotient of systolic and diastolic pressure, which are obtained by separating the daytime and nighttime blood pressure readings.

Chapter 4
Study Cases to Test the Optimization Performed in the Hybrid Model

In this chapter, each of the optimizations made to the proposed model is described and comparisons of the different algorithms used for the mentioned optimizations are made to analyze the performance of each one and determine with which one a better result is obtained, and these experiments are organized as study cases.

4.1 Optimization of the Fuzzy System to Provide the Correct Classification of the Nocturnal Blood Pressure Profile

For this first study case, the optimization of a fuzzy system is carried out to obtain the correct classification of the nocturnal blood pressure profile (NBBP), of which the methodology carried out is explained as follows.

4.1.1 Design of a Fuzzy System for Classification of Nocturnal Blood Pressure Profile

A Mamdani fuzzy inference system (FIS) is created, taking into account the experience of the expert and [1]. This FIS uses two inputs, which are the quotient of the systolic pressure and the quotient of the diastolic pressure, the range goes from 0.4 to 1.3 and is granulated into four membership functions using "GreaterFall", "Fall", "Increase" and "GreaterIncrease" as linguistic variables. The output represents the nocturnal blood pressure profile, which uses the range from 0 to 100% and is granulated in four using "Extreme Dipper", "Dipper", "Non-Dipper" and "Riser" as linguistic variables. Figure 4.1 represents the proposed fuzzy system.

© The Author(s), under exclusive license to Springer Nature Switzerland AG 2022
P. Melin et al., *Nature-inspired Optimization of Type-2 Fuzzy Neural Hybrid Models for Classification in Medical Diagnosis*, SpringerBriefs in Computational Intelligence, https://doi.org/10.1007/978-3-030-82219-4_4

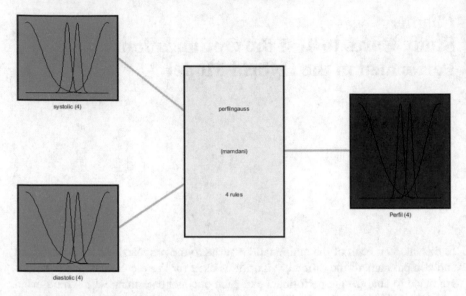

Fig. 4.1 Nocturnal blood pressure profile fuzzy system

Two versions of this fuzzy system are designed, the first using Trapezoidal membership functions and the second using Gaussian membership functions to make a comparison of the generated results. Figure 4.2 presents the inputs and output of the second version of the fuzzy system designed.

The rules used for the fuzzy system are four and are presented in Table 4.1

To perform the optimization, the CSO and CSA algorithms are used with the objective of comparing the performance of both algorithms, as well as observing which one generates better results. The algorithms are used to generate the necessary adjustments in each point to the membership functions to obtain the one that generates the least error. To identify this as an objective is used the Mean Square Error (MSE) function, which is presented in Eq. 4.1.

$$\text{MSE} = \frac{1}{2} \sum_{i=1}^{n} \left(Y_i - \hat{Y}_i \right)^2 \tag{4.1}$$

where:

n: corresponds to the number of data points,
Y: corresponds to the observed values,
$\hat{Y}$:   corresponds to the predicted values.

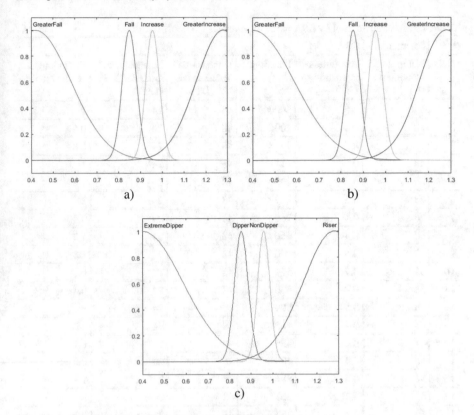

Fig. 4.2 a Systolic input, b Diastolic input c NBBP output

Table 4.1 Rules used for the fuzzy system	Systolic	Diastolic	Output
	Greater fall	Greater fall	Extreme dipper
	Fall	Fall	Dipper
	Increase	Increase	Non-dipper
	Greater increase	Greater increase	Riser

4.1.2 Experimentation and Results

With the CSA algorithm, the first 30 experiments are carried out, varying their parameters, which are presented in Table 4.2. Such optimization is done for both fuzzy systems created with Trapezoidal and Gaussian membership functions.

For these experiments the awareness probability (AP) was varied between 0 and 1 s and the flight length (Fl) is a random number between 0 and 2.

Table 4.2 Parameters used by CSA in the experiments

	Parameters of CSA					
No	Number of individuals (N)	Awareness Probability (AP)	Iterations (Iter)	Dimensions Trapezoidal MF (DT)	Dimensions Guassian MF (DGT)	Flight length (FL)
1	5	0.5	4000	48	24	2
2	10	0.3	2000	48	24	0.5
3	15	0.7	1333	48	24	0.8
4	18	0.9	1111	48	24	1
5	20	0.2	1000	48	24	1.5
6	23	0.8	869	48	24	2
7	26	0.1	769	48	24	0.4
8	30	0.4	666	48	24	1.3
9	33	0.6	606	48	24	1.8
10	35	0.7	571	48	24	0.7
11	38	0.8	526	48	24	0.3
12	40	0.5	500	48	24	1
13	43	0.3	465	48	24	1.2
14	47	0.2	425	48	24	0.1
15	50	0.4	400	48	24	0.9
16	52	0.6	384	48	24	1.7
17	55	0.2	363	48	24	2
18	60	0.1	333	48	24	0.3
19	64	0.9	312	48	24	0.6
20	67	0.7	298	48	24	1.6
21	70	0.4	285	48	24	1.4
22	72	0.5	277	48	24	0.2
23	75	0.3	266	48	24	0.7
24	80	0.4	250	48	24	1.1
25	85	0.2	235	48	24	2
26	87	0.6	229	48	24	0.5
27	90	0.1	222	48	24	0.8
28	92	0.8	217	48	24	1
29	96	0.9	208	48	24	0.4
30	100	0.5	200	48	24	0.7

Thirty experiments are performed with the CSO algorithm to observe its performance. In each experiment, each of the parameters was varied, which are listed in Table 4.3, where:

Pop is the population, *M* is the iterations, *DT* is the dimension used for the fuzzy system with trapezoidal membership functions, *DG* is the dimension used for the fuzzy system with Gaussian membership functions, *G* is how often the chicken

Table 4.3 Parameters used for CSO in the experiments

CSO parameters								
N.E	Pop	M	DT	DG	G	r%	h%	m%
1	5	4000	48	24	14	0.15	0.6	0.3
2	10	2000	48	24	9	0.15	0.8	0.5
3	15	1333	48	24	3	0.15	0.7	0.3
4	18	1111	48	24	15	0.15	0.6	0.4
5	20	1000	48	24	16	0.15	0.5	0.2
6	23	869	48	24	20	0.15	0.7	0.1
7	26	769	48	24	15	0.15	0.8	0.5
8	30	666	48	24	16	0.15	0.6	0.2
9	33	606	48	24	4	0.15	0.5	0.5
10	35	571	48	24	14	0.15	0.8	0.1
11	38	526	48	24	9	0.15	0.6	0.4
12	40	500	48	24	9	0.15	0.7	0.3
13	43	465	48	24	16	0.15	0.8	0.2
14	47	425	48	24	12	0.15	0.5	0.4
15	50	400	48	24	9	0.15	0.6	0.4
16	52	384	48	24	16	0.15	0.6	0.5
17	55	363	48	24	3	0.15	0.7	0.2
18	60	333	48	24	8	0.15	0.8	0.4
19	64	312	48	24	11	0.15	0.5	0.1
20	67	298	48	24	2	0.15	0.7	0.1
21	70	285	48	24	5	0.15	0.6	0.4
22	72	277	48	24	17	0.15	0.8	0.2
23	75	266	48	24	7	0.15	0.7	0.1
24	80	250	48	24	6	0.15	0.6	0.4
25	85	235	48	24	19	0.15	0.8	0.3
26	87	229	48	24	4	0.15	0.6	0.3
27	90	222	48	24	9	0.15	0.7	0.1
28	92	217	48	24	19	0.15	0.8	0.5
29	96	208	48	24	13	0.15	0.6	0.1
30	100	200	48	24	10	0.15	0.5	0.3

swarm can be updated, $r\%$ *is* the population size of roosters, $h\%$ is the population size of hens and $m\%$ is the population size of mother hens.

80 patients are used to test the fuzzy system generated by both algorithms, and with them to be able to compare the results obtained and determine which algorithm provides the best results. Table 4.4 listed the percentage of classification obtained in the experiments performed by the CSA and CSO algorithms respectively.

Table 4.4 Comparison of percent of success obtained by the algorithms

% success Trapezoidal MFs		% success Gaussian MFs	
CSO	CSA	CSO	CSA
92.5	86.25	85	72.25
90	*81.25*	83.75	*72.5*
88.75	87.5	83.75	95
91.25	87.5	85	78.75
87.5	81.25	85	88.75
90	88.75	90	86.25
92.5	88.75	87.5	75
91.25	86.25	90	83.75
91.25	86.25	91.25	91.25
91.25	86.25	86.25	76.25
90	87.5	**92.5**	78.75
90	88.75	87.5	90
91.25	88.75	87.5	80
87.5	87.5	88.75	80
90	86.25	88.75	92.5
93.75	88.75	81.25	76.25
91.25	87.5	87.5	87.5
93.75	87.5	87.5	88.75
91.25	87.5	86.25	90
95	87.5	86.25	83.75
93.75	88.75	88.75	83.75
93.75	88.75	88.75	78.75
91.25	86.25	88.75	75
93.75	87.5	90	88.75
90	88.75	87.5	**95**
92.5	88.75	90	91.25
91.25	87.5	88.75	92.5
92.5	86.25	87.5	90
90	88.75	87.5	85
95	88.75	86.25	93.75

Table 4.5 Summary of the average

CSO		CSA	
Trapezoidal MFs (%)	Gaussian MFs (%)	Trapezoidal MFs (%)	Gaussian MFs (%)
91.46	87.59	87.25	84.7

Analyzing each of the experiments carried out by both algorithms, it can be observed that in the optimization with the CSO algorithm using Trapezoidal membership functions, the highest classification was 95% in experiment 21, while with the CSA good results were not obtained. For the case of the fuzzy system with Gaussian membership functions, with the CSO the highest classification was 92.5%, while with the CSA it was 95% in experiment 25 and in general the lowest classification obtained was with this same fuzzy and with only 72.25% in experiments 1 and 2.

The average obtained from the 30 experiments carried out by both algorithms is presented in Table 4.5. It is observed that in the two FISs designed the CSO algorithm provides a higher average classification. Analyzing the results of the fuzzy system using trapezoidal membership functions, it is observed that 91.46% of classification is obtained, whereas for the fuzzy system using Gaussian membership functions is obtained on average of 87.25% correct classification.

4.1.3 Statistical Test

To determine which of the used algorithms provided the least error, a statistical test is carried out using the parametric Z test. The experimentation carried out is revealed in two cases, beginning to study the results obtained with the fuzzy systems designed with Trapezoidal membership functions. The results of the 30 experiments generated by the CSA and CSO algorithms are studied, from which the averages of the classification provided are taken.

The formula for the Z test is expressed mathematically as follows:

$$Z = \frac{(\bar{x}_1 - \bar{x}_2) - (\mu_1 - \mu_2)}{\sigma_{\bar{x}_1 - \bar{x}_2}} \tag{4.2}$$

where:

$\bar{x}_1 - \bar{x}_2$:   It is the observed difference.

$\mu_1 - \mu_2$:   It is the expected difference.

$\sigma_{\bar{x}_1 - \bar{x}_2}$:   Standard error of the differences.

It is established as a null hypothesis that the average of the classification obtained by the CSO algorithm is less than or equal to the average of the classification obtained by the CSA algorithm. The alternative hypothesis establishes that the average obtained derived from the classification of the CSO algorithm is greater

Table 4.6 Hypothesis test parameters

Parameters	
C. interval	95%
alpha	0.05
Ho	$\mu_1 \leq \mu_2$
Ha	$\mu_1 > \mu_2$
C. value	$Z = 1.645$

Table 4.7 Z-test descriptive statistics

Variable	Observations	Mean	Std. deviation
CSO	30	91.458	1.944
CSA	30	87.250	1.897

Table 4.8 Z-test results

Difference	4.208
z (Observed value)	8.485
z (Critical value)	1.645
p-value	0
alpha	0.05

than the average obtained from the experiments carried out by the CSA algorithm. In Table 4.6 the parameters of the Z test are presented.

The descriptive statistics of the first case study are presented in Table 4.7.

The results obtained derived from the Z test are presented in Table 4.8.

From the results obtained, the following is concluded: Because the result of the p-value is less than the level of significance alpha $= 0.05$, the null hypothesis is rejected and the alternative hypothesis is accepted, with which it can be determined that there is sufficient evidence with a 5% significance level to support the claim that the ranking provided by the CSO algorithm is higher than the ranking generated by the CSA algorithm.

For the second case, 30 experiments are performed using the CSO and CSA algorithms for optimization of the fuzzy system using Gaussian membership functions, from which the average of classification obtained by the aforementioned methods is used to enhance the statistical test.

The null hypothesis is established that the average of the classification obtained by the CSO algorithm is less than or equal to the average of the classification obtained by the CSA algorithm. The alternative hypothesis establishes that the average obtained derived from the classification of the CSO algorithm is greater than the average obtained from the experiments carried out by the CSA algorithm. In Table 4.9 the Z-Test parameters are presented.

The descriptive statistics of the second case study are presented in Table 4.10.

The results obtained derived from the Z test are presented in Table 4.11.

Table 4.9 Hypothesis test parameters

Parameters	
C. interval	95%
alpha	0.05
Ho	$\mu_1 \leq \mu_2$
Ha	$\mu_1 > \mu_2$
C. value	$Z = 1.645$

Table 4.10 Descriptive statistics of Z-test

Variable	Observations	Mean	Std. deviation
CSO	30	87.500	2.390
CSA	30	84.700	7.030

Table 4.11 Z-test results

Difference	2.800
z (Observed value)	2.065
z (Critical value)	1.645
p-value	0.0165
alpha	0.05

From the results obtained, the following is concluded: Because the result of the p-value is less than the level of significance alpha = 0.05, the null hypothesis is rejected and the alternative hypothesis is accepted, with which it can be determined that there is sufficient evidence with a 5% significance level to support the claim that the rating provided by the CSO algorithm is better than the rating generated by the CSA algorithm.

4.2 Fuzzy System Optimization to Obtain the Heart Rate Level

For this second case study, the optimization of the Type-1 and Interval Type-2 fuzzy system is performed for obtaining the correct classification of the heart rate in different patients, of which the methodology performed is explained as follows:

4.2.1 Proposed Method for Optimization of the Heart Rate Fuzzy Classifier

For this part of the proposed model, a Type-1 fuzzy system is created for obtaining the correct classification of the heart rate level. This is optimized by the Bird Swarm Algorithm to provide an accurate diagnosis. Once this is done, the parameters of the best fuzzy system provided by the algorithm are taken to design the Interval Type-2 fuzzy system to also be optimized, to compare the results of each of them and determine which one is provided the best diagnosis.

The methodology proposed for this optimization is presented in Fig. 4.3, and is carried out as follows:

From the generated database, patient information is taken to give it as input to the fuzzy system in each iteration of the algorithm and to test the classification obtained. The first input that the fuzzy system uses is the patient's age and the second input is the heart rate trend. Besides, the classifier is tested with Gaussian and Trapezoidal membership functions.

The individuals of the algorithm, which are the birds, move the parameters of the membership functions (MF) in the given search space while they find a fuzzy system that provides the smallest error in the classification. To obtain this measurement, a set of 30 test patients is used, which enters the algorithm at each iteration, and to measure the generated error, the MSE is used as the objective function, and this is presented in Eq. 4.1.

Since it has the Type-1 fuzzy system that has provided the best classification, the parameters of the membership functions are used as a base to construct the Interval Type-2 fuzzy system, of which is optimized the footprint of uncertainty.

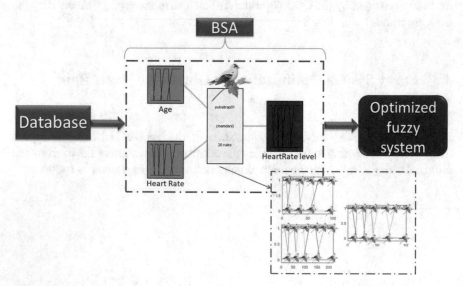

Fig. 4.3 Model used for the optimization of Heart Rate fuzzy classifier

4.2.1.1 Design of Type-1 Fuzzy System for Obtaining the Heart Rate Level Using Trapezoidal MF

The fuzzy system designed to solve this problem is of the Mamdani type, based on the expert's experience and [2], which has two inputs which corresponds to the age and the trend of the heart rate and the output corresponds to the heart rate level. This is presented in Fig. 4.4 and described as follows:

The first input, that is the age is presented in Fig. 4.5 and is grainy with four MFs, which use "Child", "Young", "Adult", and "Elder" as linguistic variables, in addition to using a range from 0 to 100.

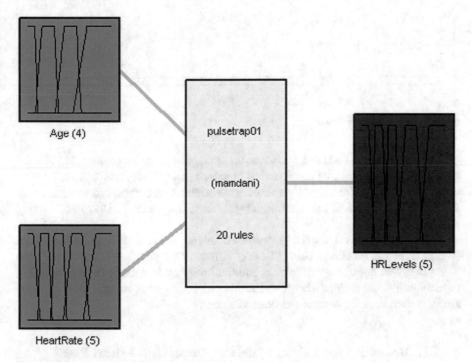

Fig. 4.4 Heart Rate Level Fuzzy System

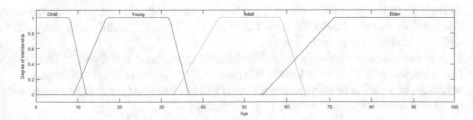

Fig. 4.5 Age input

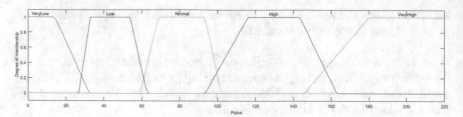

Fig. 4.6 Heart rate input

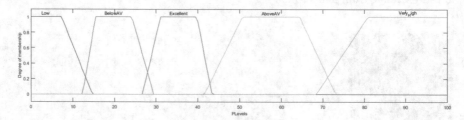

Fig. 4.7 Heart rate level output

The second input, that corresponds to the trend of heart rate is presented in Fig. 4.6 and is granulated with five membership functions, using "VeryLow", "Low", "Normal", "High" and "VeryHigh" as linguistic variables, in addition to using a range from 0 to 220, since it is the maximum heart rate that a person can have while doing physical activity.

The heart rate level output is presented in Fig. 4.7 which is granulated with five membership functions, using "Low", "BelowAV", "Excellent", "AboveAV" and "VeryHigh" as linguistic variable, in addition to going in a range from 0 to 100%, determining if the patient's heart rate level is high, low or at an optimal level. This fuzzy system uses the centroid as defuzzification.

4.2.1.2   Design of Type-1 Fuzzy System for Obtaining the Heart Rate Level Using Gaussian MF

The age input is presented in Fig. 4.8 and is granulated with four membership functions, which use the linguistic variables "Child", "Young", "Adult" and "Elder", additional using a range from 0 to 100.

The heart rate input is presented in Fig. 4.9 and is granulated with five membership functions, using "VeryLow", "Low", "Normal", "High" and "VeryHigh" as linguistic variables, in addition to using a range from 0 to 220, since it is the maximum pulse that a person can have while doing physical activity.

The heart rate level output is presented in Fig. 4.10 which is granulated with five membership functions, which corresponds to "Low", "BelowAV", "Excellent", "AboveAV" and "VeryHigh" linguistic variables, in addition to that it goes in a range

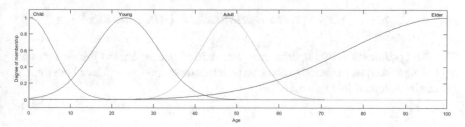

Fig. 4.8 Age input

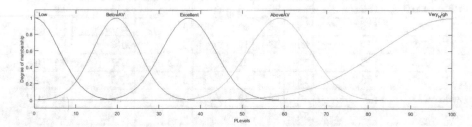

Fig. 4.9 Heart rate input

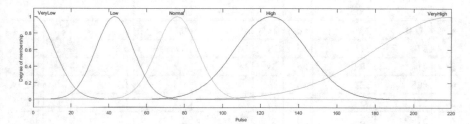

Fig. 4.10 Heart rate level output

from 0 to 100%, determining if the patient's heart rate level is high, low or at an optimal level. This fuzzy system uses the centroid as defuzzification.

Table 4.12 presents the fuzzy rules, these are based on the age and tendency of the heart rate to determine the classification of the heart rate level.

Table 4.12 Rules of the heart rate level FIS

Age/Heart rate	Very Low	Low	Normal	High	Very High
Child	Low	Low	Excellent	Excellent	AboveAV
Young	Low	BelowAV	Excellent	AboveAV	VeryHigh
Adult	Low	BelowAV	Excellent	AboveAV	VeryHigh
Elder	Low	BelowAV	Excellent	VeryHigh	VeryHigh

4.2.2   Type-1 Fuzzy System Optimization Using the BSA

For this optimization, 30 experiments were carried out, where the parameters used by the algorithm were varied to observe which combination generates a better result, which is presented in Table 4.13.

Table 4.13 BSA Combination of parameters in the experiments

No	Iterations (M)	Population (pop)	DT	DG	Frequency (FQ)	Cognitive (c1)	Social (c2)	a1	a2
1	1000	20	56	28	19	0.5	0.5	2	2
2	870	24	56	28	28	0.8	0.8	1.5	1.5
3	714	28	56	28	18	1.2	1.2	0.4	0.4
4	625	32	56	28	15	1.5	1.5	0.1	0.1
5	571	36	56	28	6	1.8	1.8	0.8	0.8
6	574	38	56	28	21	2	2	1	1
7	454	44	56	28	25	2.33	2.33	1.3	1.3
8	416	48	56	28	6	2.48	2.48	0.6	0.6
9	400	50	56	28	28	2.76	2.76	0.9	0.9
10	357	56	56	28	20	3	3	1.1	1.1
11	338	60	56	28	10	3.18	3.18	1.9	1.9
12	322	62	56	28	21	3.22	3.22	0.5	0.5
13	307	66	56	28	1	3.45	3.45	1.5	1.5
14	285	70	56	28	13	3.56	3.56	0.7	0.7
15	278	72	56	28	2	4	4	1.3	1.3
16	256	78	56	28	24	0.4	0.4	1.8	1.8
17	250	80	56	28	19	0.7	0.7	0.3	0.3
18	235	86	56	28	1	1.15	1.15	0.9	0.9
19	227	88	56	28	24	1.34	1.34	1	1
20	208	96	56	28	22	1.45	1.45	2	2
21	202	100	56	28	15	1.67	1.67	0.6	0.6
22	166	120	56	28	4	1.78	1.78	0.3	0.3
23	133	150	56	28	16	1.92	1.92	1.5	1.5
24	111	180	56	28	2	2.18	2.18	1.2	1.2
25	100	200	56	28	2	2.39	2.39	1.8	1.8
26	95	210	56	28	21	2.56	2.56	0.7	0.7
27	90	220	56	28	22	2.83	2.83	0.9	0.9
28	87	230	56	28	15	3.4	3.4	1.5	1.5
29	83	240	56	28	19	3.7	3.7	1.7	1.7
30	80	250	56	28	20	4	4	2	2

Input Age			
Child	Young	Adult	Elder
1 2 3 4	5 6 7 8	9 10 11 12 13	14 15 16

Input Pulse				
VeryLow	Low	Normal	High	VeryHigh
17 18 19 20	21 22 23 24	25 26 27 28	29 30 31 32	33 34 35 36

Output PLevels				
Low	BelowAV	Excellent	AboveAV	VeryHigh
37 38 39 40	41 42 43 44	45 46 47 48	49 50 51 52	53 54 55 56

Fig. 4.11 BSA individual representation in FIS with trapezoidal MFs

Input Age				Input Pulse				
Child	Young	Adult	Elder	VeryLow	Low	Normal	High	VeryHigh
1	2	3	4	5	6	7	8	9	10	11	12	13	14	15	16	17	18

Output PLevels				
Low	BelowAV	Excellent	AboveAV	VeryHigh
19	20	21	22	23	24	25	26	27	28

Fig. 4.12 BSA individual representation in FIS with Gaussians MFs

Trapezoidal membership functions representation is illustrated in Fig. 4.11. In the same way, the Gaussian membership functions representation is presented in Fig. 4.12. The numbers in both figures represent the points in the parameters of the membership functions in the fuzzy systems, that are adjusted by the algorithm.

4.2.3   Design and Optimization of the IT2FS

The best points of the Type-1 fuzzy system generated by the BSA are used as a reference for construct of the interval Type-2 fuzzy system, leaving the footprint of uncertainty symmetrically.

The inputs of the IT2FS using Trapezoidal MFs are illustrated in Figs. 4.13 and 4.14 respectively, where the output is illustrated in Fig. 4.15.

The inputs of the IT2FS with Gaussian membership functions are represented in Figs. 4.16 and 4.17 respectively, while the output is represented in Fig. 4.18.

As mentioned above, the footprint of uncertainty of the IT2FS was optimized to obtain the ideal aperture and thereby provide an accurate classification. The representation of the parameters optimized by the algorithm for the fuzzy system using Gaussian MFs is Illustrated in Fig. 4.19. The representation of the IT2FS using trapezoidal MFs can be illustrated as the Type-1 fuzzy system with Fig. 4.11 because when respecting the optimized points of Type-1 fuzzy system, the same number of

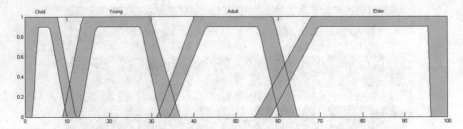

Fig. 4.13 Age input using trapezoidal MFs

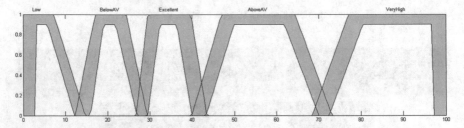

Fig. 4.14 Heart rate input using trapezoidal MFs

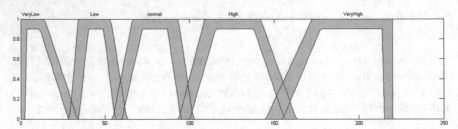

Fig. 4.15 Heart rate level output using trapezoidal MFs

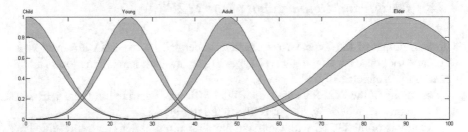

Fig. 4.16 Age input using Gaussian MFs

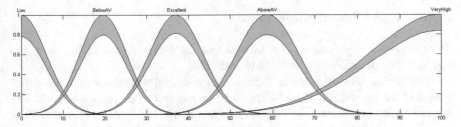

Fig. 4.17 Heart rate input using Gaussian MFs

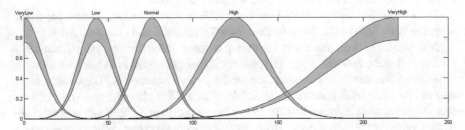

Fig. 4.18 Heart rate level using Gaussian MFs

Input Age				Input Pulse				
Child	Young	Adult	Elder	VeryLow	Low	Normal	High	VeryHigh
1 2 3	4 5 6	7 8 9	10 11 12	13 14 15	16 17 18	19 20 21	22 23 24	25 26 27

Output Plevels				
Low	BelowAV	Excellent	AboveAV	VeryHigh
28 29 30	31 32 33	34 35 36	37 38 39	40 41 42

Fig. 4.19 BSA representation for the IT2FS using Gaussians MFs

parameters remains to be used. For this Fuzzy System, the rules presented in Table 4.12 are used.

4.2.4 Results Obtained from Optimizing the Heart Rate Fuzzy System

Next, the obtained results derived from the different optimizations carried out are presented:

4.2.4.1   Type-1 Fuzzy Systems Results

30 experiments were carried out to optimize the Fuzzy Systems, in which the parameters of the BSA were varying in each case.

These parameters are listed in Table 4.13, where M are the iterations, *pop* corresponds to the population of birds, *DimT* is the number of dimensions for the FIS that use trapezoidal MFs, *DimG* the dimensions for the FIS that use Gaussian MFs, *FQ (FL)* represents the frequency of bird behavior, *c1* is the cognitive accelerated coefficient, *c2* is the social accelerated coefficient, *a1* and *a2* are parameters related to the indirect and direct effect of bird vigilance behavior.

The parameters that generated the best result are presented in Table 4.14, which were the best, both in the fuzzy system with Trapezoidal and Gaussian MFs.

It is interesting and extremely important to know the classification percentage in each experiment, in such a way that these results are presented in Table 4.15. Column 3 describes the results of classification for the fuzzy systems with Trapezoidal MFs, where it is observed that in experiments 14 and 28 a 100% classification was obtained and the worst classification was 87.5% in the experiment 8. While in column 3 the results of classification for the fuzzy systems with Gaussian MFs are presented, observing that experiment 14 was the one that generated a better classification with 95%, while the worst classification was also 87.5% in several of the performed experiments. The best experiments are highlighted in bold type.

In summary, the average classification obtained from the 30 experiments is as follows:

Trapezoidal MFs: 94.58%
Gaussian MFs: 90.33%

The input of the FIS with optimized Trapezoidal MFs are illustrated in Figs. 4.20 and 4.21. Likewise, the output of the same fuzzy system is illustrated in Fig. 4.22.

The inputs of the fuzzy system with optimized Gaussian membership functions are presented in Figs. 4.23 and 4.24. While the output of the mentioned fuzzy system is presented in Fig. 4.25. In both cases, the adjustment made by the BSA algorithm in each membership function can be observed.

Table 4.14 Summary of best parameters obtained

Iterations	285
Population	70
DimT	56
DimG	28
F. of behavior	13
Cognitive A.C	3.56
Social A.C	3.56
a1	0.7
a2	0.7

Table 4.15 Experiment with BSA that represents the percentage of classification

No	TrapMFs (%)	GaussMFs (%)
1	97.50	87.50
2	95	87.50
3	97.50	92.50
4	90	87.50
5	97.50	87.50
6	90	92.50
7	92.50	90
8	*87.50*	92.50
9	92.50	92.50
10	95	90
11	95	92.50
12	90	95
13	92.50	90
14	**100**	**95**
15	97.50	90
16	95	90
17	95	87.50
18	97.50	90
19	92.50	92.50
20	95	92.50
21	95	92.50
22	92.50	85
23	95	90
24	95	87.50
25	100	90
26	95	87.50
27	97.50	92.50
28	**100**	87.50
29	92.50	90
30	90	92.50

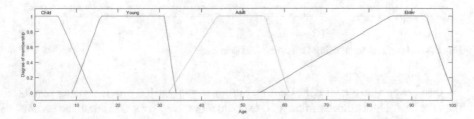

Fig. 4.20 Input age optimized using BSA

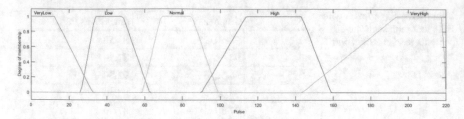

Fig. 4.21 Input heart rate optimized using BSA

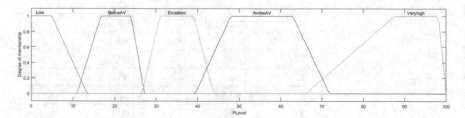

Fig. 4.22 Heart rate level optimized using BSA

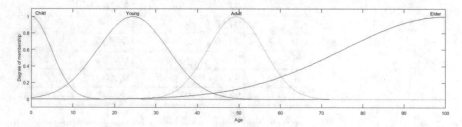

Fig. 4.23 Age input using Gaussian MFs optimized

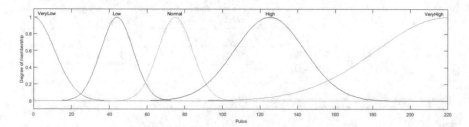

Fig. 4.24 Heart rate input using Gaussian MFs optimized

Tests are carried out with the CSA algorithm, to analyze if it provides better results than the BSA algorithm. Similarly, 30 experiments were performed taken the parameters listed in Table 4.13. The percentage of classification generated by each fuzzy system is obtained, which is presented in Table 4.16, where it is observed that for the

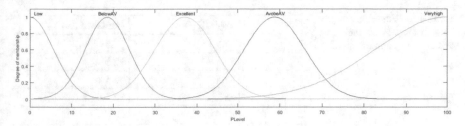

Fig. 4.25 Heart rate level output using Gaussian MF optimized

classification of fuzzy systems with Trapezoidal MFs, which is found in column 2, it was 97.5% in experiment 22, while for the classification with Gaussian MFs, which are found in column 3, it was 92.5% in experiment 21. For the optimization of both MFs, the worst classification was 87.5% in several of the experiments performed.

As an average of the 30 experiments, the following was obtained:

Trapezoidal MFs: 88.33%

Gaussian MFs: 89.66%

The comparison of the averages obtained from the optimization by CSA and BSA are presented in Table 4.17, it is observed that the highest percentage of classification is contained by the algorithm BSA for the two membership functions used.

4.2.4.2   Optimization of the IT2FS

In the same way that the optimization was realized in the Type-1 fuzzy system, for the IT2FS, 30 different experiments were performed, using the parameters presented in Table 4.13. The best classification was given by the fuzzy system with Trapezoidal MFs generated in Experiment 21 with a 97.5% classification rate. Table 4.18 presents the summary of the best parameters obtained.

For the optimization of the IT2FS with Gaussian membership functions, a 100% classification was obtained with experiment 28. The parameters of this experiment are presented in Table 4.19.

The classification percentages of the 30 experiments for both IT2FS are presented in Table 4.20. On the contrary, with was observed in the optimization of Type-1 FS, for IT2FS using Trapezoidal MFs, the best classification obtained was 97.4% in experiment 21, while optimizing the Gaussian MFs 100% classification was obtained in experiment 28.

For the IT2FS using Trapezoidal membership functions, the average classification was 92.92%, whereas for the IT2FS was obtained an average of 92.78% classification.

The optimized footprint of uncertainty is presented as follows:

In Figs. 4.26 and 4.27 the inputs are presented, likewise, the output is illustrated in Fig. 4.28.

The optimized Gaussian membership functions are presented as follows:

Table 4.16 Experiment with CSA that represents the percentage of classification

No	TrapMF (%)	GaussMF (%)
1	87.5	87.5
2	87.5	92.5
3	87.5	92.5
4	85	87.5
5	87.5	90
6	85	90
7	87.5	90
8	87.5	90
9	87.5	90
10	85	90
11	90	87.5
12	90	90
13	90	90
14	90	90
15	87.5	90
16	85	90
17	87.5	90
18	90	90
19	87.5	87.5
20	90	87.5
21	90	**92.5**
22	**97.5**	90
23	90	90
24	90	90
25	87.5	87.5
26	87.5	87.5
27	90	90
28	87.5	90
29	87.5	90
30	87.5	90

Table 4.17 BSA and CSA comparison of averages

Trapezoidal MFs		Gaussians MFs	
BSA (%)	CSA(%)	BSA (%)	CSA (%)
94.58	88.33	90.33	89.66

Table 4.18 Summary of the best parameters obtained for Trapezoidal MFs

Iterations	202
Population	100
DimT	56
F. of the behavior	15
Cognitive A.C	1.67
Social A.C	1.67
a1	0.6
a2	0.6

Table 4.19 Summary of the best parameters obtained for Gaussian MFs

Iterations	87
Population	230
DimG	28
F. of the behavior	15
Cognitive A.C	3.4
Social A.C	3.4
a1	1.5
a2	1.5

Both inputs are illustrated in Figs. 4.29 and 4.30, whereas the output is illustrated in Fig. 4.31. It can be observed that the adjustment in the footprint of uncertainty is noticeable.

The CSA algorithm is used to compare whether it generates better results than the BSA algorithm, for this, 30 experiments were realized, using also the parameters of Table 4.13. The classification of each of the generated IT2FS is obtained, which are presented in Table 4.21. Where it can be analyzed that the best classification is provided in experiment 1 with 92.50%, for the Trapezoidal MFs, while the best classification for the fuzzy systems with Gaussian MFs was 95%.

The average classification in fuzzy systems that used trapezoidal MFs was 92.83%, whereas for fuzzy systems with Gaussian MFs it was 88.33%. Table 4.22 presents the comparison of the averages obtained by each algorithm for both membership functions, and it can be determined that in both cases a higher average is obtained with the BSA algorithm.

4.2.4.3 Tests Performed with the Type-1 Fuzzy Systems

To test the FIS, 15 random patients are used, since the patients in the database generated with the patient group do not have much variation in both entries, and it is necessary to ensure that the classification is being carried out correctly. Analyzing the information, it is observed that the classification is carried out correctly.

No	TrapMF (%)	GaussMF (%)
Table 4.20 Experiment with CSA that represents the percentage of classification		
1	92.50	92.50
2	92.50	90
3	92.50	87.50
4	95	90
5	92.50	87.50
6	92.50	92.50
7	92.50	95
8	92.50	90
9	92.50	92.50
10	92.50	95
11	92.50	97.50
12	92.50	87.50
13	92.50	95
14	92.50	92.50
15	95	92.50
16	92.50	85
17	95	92.50
18	92.50	92.50
19	92.50	92.50
20	95	97.50
21	**97.50**	90
22	87.50	97.50
23	92.50	100
24	92.50	92.50
25	92.50	87.50
26	92.50	90
27	92.50	97.50
28	92.50	**100**
29	92.50	97.50
30	95	92.50

The patient data used by the fuzzy systems are presented in columns 2 and 3 of Table 4.23. Likewise, it is observed in columns 5 to column 8 the results of the optimized and non-optimized fuzzy systems are presented. Analyzing the results of the Trapezoidal fuzzy systems, with the non-optimized FIS classifies 13 patients in correct form, whereas the optimized FIS classifies 15 patients in correct form. For Gaussian membership functions, the non-optimized fuzzy system classifies 12 patients in correct form, while the optimized fuzzy system classifies 14 patients in correct form. Incorrectly classified patients are written in italics. Fuzzy systems

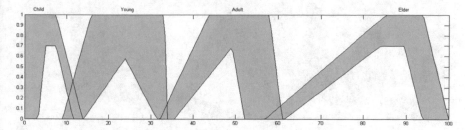

Fig. 4.26 Age input sing trapezoidal MFs optimized for BSA

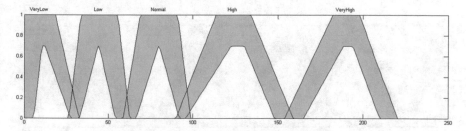

Fig. 4.27 Heart rate input using trapezoidal MFs optimized for BSA

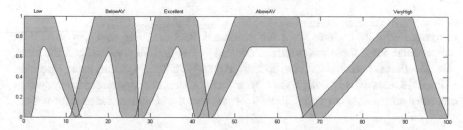

Fig. 4.28 Heart rate level output using trapezoidal MFs optimized for BSA

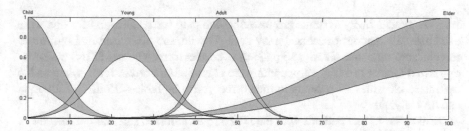

Fig. 4.29 Age input using Gaussian MFs optimized for BSA

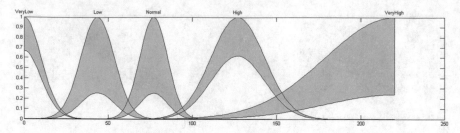

Fig. 4.30 Heart rate input using Gaussian MFs optimized for BSA

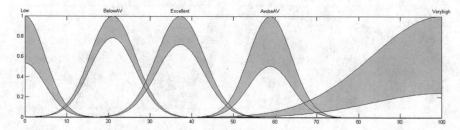

Fig. 4.31 Heart rate level output using Gaussian MFs optimized for BSA

are tested with 20 patients from the generated database, with fuzzy systems with Trapezoidal MFs, the non-optimized correctly classified 19 patients, and the same classification is obtained with the optimized one. With fuzzy systems using Gaussian MFs, it also classifies 19 patients in correct form, whereas the optimized fuzzy system classifies all patients correctly. Table 4.24 presents the aforementioned information.

4.2.4.4   Tests Performed Using IT2FS

Interval Type-2 fuzzy systems are tested to compare the results, and for this were used also, 15 random patients. Fuzzy systems with non-optimized and optimized membership functions classify 13 patients in correct form, whereas IT2FS with non-optimized Gaussian MFs, 12 patients were classified in correct form, whereas the optimized classifies 15 patients in the correct form. In Table 4.25 this information can be analyzed.

Testing the 15 real patients with the IT2FS, the optimized and non-optimized FIS with Trapezoidal MFs classifies 19 patients in correct form, as well as the optimized fuzzy system. For the fuzzy system non-optimized that use Gaussian MFs, 18 patients are classified in correct form, while with the optimized FIS, 20 patients are classified correctly. This information can be analyzed in Table 4.26.

Table 4.27 presents a summary of the classification percentages of the fuzzy systems tested in the fifteen random patients.

Table 4.21 Experiment with CSA that represents the percentage of classification

No	TrapMF (%)	GaussMF (%)
1	92.50	92.50
2	90	92.50
3	90	92.50
4	90	95
5	90	92.50
6	90	95
7	90	92.50
8	90	92.50
9	90	92.50
10	90	87.50
11	90	95
12	90	92.50
13	90	95
14	90	92.50
15	90	95
16	90	92.50
17	90	95
18	90	95
19	90	92.50
20	90	95
21	90	95
22	90	95
23	90	92.50
24	90	92.50
25	90	87.50
26	90	92.50
27	90	92.50
28	90	92.50
29	90	87.50
30	90	92.50

Table 4.22 BSA and CSA comparison of averages

Trapezoidal MFs		Gaussians MFs	
BSA (%)	CSA (%)	BSA (%)	CSA (%)
92.92	92.83	92.78	88.33

Table 4.23 Random patients using for the test

No	Age	Heart rate	Real values	Trapezoidal		Gaussian	
				Non-optimized FS	Optimized FS	Non-optimized FS	Optimized FS
1	25	84	Excellent	Excellent	Excellent	Excellent	Excellent
2	83	95	Above Avg	*Excellent*	Above Avg	Above Avg	Above Avg
3	15	114	Above Avg	Above Avg	Above Avg	Above Avg	Above Avg
4	34	72	Excellent	Excellent	Excellent	Excellent	Excellent
5	42	135	Above Avg	Above Avg	Above Avg	Above Avg	Above Avg
6	91	97	Very High	*Above Avg*	Very High	*Above Avg*	Very High
7	45	60	Below Avg	Below Avg	Below Avg	*Excellent*	*Excellent*
8	56	87	Excellent	Excellent	Excellent	*Above Avg*	Excellent
9	75	102	Very High	Very High	Very High	Very High	Very High
10	9	120	Excellent	Excellent	Excellent	Excellent	Excellent
11	14	92	Excellent	Excellent	Excellent	Excellent	Excellent
12	38	78	Excellent	Excellent	Excellent	Excellent	Excellent
13	29	80	Excellent	Excellent	Excellent	Excellent	Excellent
14	21	62	Excellent	Excellent	Excellent	Excellent	Excellent
15	6	115	Excellent	Excellent	Excellent	Excellent	Excellent

Table 4.28 presents a summary of the classification percentage of the fuzzy systems using 20 real patients.

The results generated for Type-1 and IT2FS using Gaussian MFs provides 100% of classification in both groups of patients studied. Analyzing Tables 4.27 and 4.28, it is interesting to mention that with the non-optimized fuzzy system using Gaussian MFs it obtained only 80% of correct classification.

4.3   Optimization of the Modular Neural Network to Obtain the Trend of the Blood Pressure

For this third case study, the optimization of the modular neural network is carried out to obtain the trend of blood pressure, of which the methodology is explained as follows:

Table 4.24 Real patients using for the test

No	Age	Heart rate	Real values	Trapezoidal		Gaussian	
				Non-optimized FS	Optimized FS	Non-optimized FS	Optimized FS
1	46	75	Excellent	Excellent	Excellent	Excellent	Excellent
2	28	88	Excellent	Excellent	Excellent	Excellent	Excellent
3	30	69	Excellent	Excellent	Excellent	Excellent	Excellent
4	33	59	Below Avg	Below Avg	Below Avg	*Excellent*	Below Avg
5	31	68	Excellent	Excellent	Excellent	Excellent	Excellent
6	32	71	Excellent	Excellent	Excellent	Excellent	Excellent
7	32	66	Excellent	Excellent	Excellent	Excellent	Excellent
8	27	66	Excellent	Excellent	Excellent	Excellent	Excellent
9	31	72	Excellent	Excellent	Excellent	Excellent	Excellent
10	30	76	Excellent	Excellent	Excellent	Excellent	Excellent
11	32	81	Excellent	Excellent	Excellent	Excellent	Excellent
12	28	76	Excellent	Excellent	Excellent	Excellent	Excellent
13	31	85	Excellent	Excellent	Excellent	Excellent	Excellent
14	26	85	Excellent	Excellent	Excellent	Excellent	Excellent
15	31	77	Excellent	Excellent	Excellent	Excellent	Excellent
16	29	77	Excellent	Excellent	Excellent	Excellent	Excellent
17	45	69	Excellent	Excellent	Excellent	Excellent	Excellent
18	27	63	Excellent	Excellent	Excellent	Excellent	Excellent
19	25	107	Above Avg	Above Avg	Above Avg	Above Avg	Above Avg
20	25	95	Above Avg	*Excellent*	*Excellent*	*Excellent*	Above Avg

4.3.1 Proposed Method for Optimization of the Modular Neural Network

In this study case of the research work, the optimization of the first part of the neuro-fuzzy model is performed, which corresponds to the modular neural network to obtain the blood pressure trend. The first module is given as input the systolic pressure readings, the second module the diastolic pressure readings, and the third module the heart rate readings of the patient group. This information will be trained by the modules of the modular neural network to learn its behavior and obtain the behavior trend as an output. To perform the optimization, two metaheuristics are used to test its performance and observe with which a better result is obtained, the first is the BSA algorithm, and the second the FPA algorithm. What is sought with this optimization is the architecture that generates the best result. Speaking of a specific form, it seeks to optimize the hidden layers and the number of neurons per layer. From previous experience, it is limited to two hidden layers and 30 neurons per layer, because using more causes overtraining of the modular neural network. Each of the

Table 4.25 IT2FS test using random patients

No	Age	Heart Rate	Real value	Trapezoidal		Gaussian	
				Non-optimized FS	Optimized FS	Non-optimized FS	Optimized FS
1	25	84	Excellent	Excellent	Excellent	Excellent	Excellent
2	83	95	Above Avg	Above Avg	Above Avg	Above Avg	Above Avg
3	15	114	Above Avg	Above Avg	Above Avg	Above Avg	Above Avg
4	34	72	Excellent	Excellent	Excellent	Excellent	Excellent
5	42	135	Above Avg	Above Avg	Above Avg	Above Avg	Above Avg
6	91	97	Very High	*Above Avg*	Very High	*Above Avg*	Very High
7	45	60	Below Avg	*Excellent*	*Excellent*	*Excellent*	Below Avg
8	56	87	Excellent	Excellent	Excellent	*Above Avg*	Excellent
9	75	102	Very High	Very High	Very High	Very High	Very High
10	9	120	Excellent	Excellent	Excellent	Excellent	Excellent
11	14	92	Excellent	Excellent	*Above Avg*	Excellent	Excellent
12	38	78	Excellent	Excellent	Excellent	Excellent	Excellent
13	29	80	Excellent	Excellent	Excellent	Excellent	Excellent
14	21	62	Excellent	Excellent	Excellent	Excellent	Excellent
15	6	115	Excellent	Excellent	Excellent	Excellent	Excellent

metaheuristics used provides the architecture with which the lowest error has been obtained, for this the Mean Square Error is used as an objective function, which can be found in Eq. 4.1.

4.3.2   Results of the Optimization Made to the Modular Neural Network

This section presents the results of the optimization of the modular neural network using both metaheuristics.

4.3.2.1   Optimization of Modular Neural Network Using BSA

30 experiments are carried out to test this first metaheuristic, where parameter variation was carried out in each experiment, which is presented in Table 4.29, and which were described in Sect. 4.2.

First, the optimization of the module corresponding to the systolic pressure is carried out. Table 4.30 presents the variations of the architecture made by the BSA in

Table 4.26 IT2FS test for real patients

No	Age	Heart rate	Real values	Trapezoidal		Gaussian	
				No optimized FS	Optimized FS	No optimized FS	Optimized FS
1	46	75	Excellent	Excellent	Excellent	Excellent	Excellent
2	28	88	Excellent	Excellent	Excellent	Excellent	Excellent
3	30	69	Excellent	Excellent	Excellent	Excellent	Excellent
4	33	59	Below Avg	*Excellent*	Below Avg	*Excellent*	Below Avg
5	31	68	Excellent	Excellent	Excellent	Excellent	Excellent
6	32	71	Excellent	Excellent	Excellent	Excellent	Excellent
7	32	66	Excellent	Excellent	Excellent	Excellent	Excellent
8	27	66	Excellent	Excellent	Excellent	Excellent	Excellent
9	31	72	Excellent	Excellent	Excellent	Excellent	Excellent
10	30	76	Excellent	Excellent	Excellent	Excellent	Excellent
11	32	81	Excellent	Excellent	Excellent	Excellent	Excellent
12	28	76	Excellent	Excellent	Excellent	Excellent	Excellent
13	31	85	Excellent	Excellent	Excellent	Excellent	Excellent
14	26	85	Excellent	Excellent	Excellent	Excellent	Excellent
15	31	77	Excellent	Excellent	Excellent	Excellent	Excellent
16	29	77	Excellent	Excellent	Excellent	Excellent	Excellent
17	45	69	Excellent	Excellent	Excellent	Excellent	Excellent
18	27	63	Excellent	Excellent	Excellent	Excellent	Excellent
19	25	107	Above Avg	Above Avg	Above Avg	Above Avg	Above Avg
20	25	95	Above Avg	Above Avg	Above Avg	*Excellent*	Above Avg

the hidden layers and the number of neurons per each hidden layer, in the same way, the error obtained from performing each experiment is presented in the last column.

It can be analyzed that the best result was given in experiment 17 with an average error of 0.49 which is highlighted in bold, while the worst experiment was the number 15 with a mean error of 6.53 and which is highlighted with italics.

The second experiment carried out corresponds to the optimization of the module that generates the trend of diastolic pressure, in which variation is made in the parameters in each experiment, using those presented in Table 4.29. The results obtained are presented in Table 4.31, where it can be analyzed that the best experiment was 29 with an error of 0.003 which is highlighted in bold, while the worst was experiment 8 with a generated error of 0.131 which is highlighted in italics.

The last experiment carried out corresponds to the optimization of the module to obtain the heart rate trend, in which variation is made in the parameters in each experiment, using those presented in Table 4.29. The results obtained are presented in Table 4.32, where it can be analyzed that the best experiment was 19 with an error

Table 4.27 Summary of classification with Type-1 and IT2FS using random patients

Non-optimized T1FS		Optimized T1FS		Non-optimized IT2FS		Optimized IT2FS	
Trapezoidal	Gaussian	Trapezoidal	Gaussian	Trapezoidal	Gaussian	Trapezoidal	Gaussian
86.6%	80%	100%	93.3%	86.6%	80%	86.6%	100%

Table 4.28 Summary of classification with Type-1 and IT2FS using random patients

Non-optimized T1FS	Optimized T1FS		Non-optimized IT2FS		Optimized IT2FS	
Trapezoidal	Gaussian	Trapezoidal	Gaussian	Trapezoidal	Gaussian	Trapezoidal	Gaussian
95%	90%	95%	100%	95%	90%	100%	100%

Table 4.29 BSA parameters for each experiment

Parameters

N.E	Population (pop)	Iterations (M)	a1	a2	Frequency (FQ)	Cognitive (c1)	Social (c2)	Dimen (Dim)
1	10	800	0.8	0.8	2	0.3	0.3	4
2	12	667	0.3	0.3	11	1.4	1.4	4
3	14	571	1.2	1.2	6	2	2	4
4	15	533	1.9	1.9	14	4.3	4.3	4
5	18	444	2	2	14	0.1	0.1	4
6	20	400	0.1	0.1	13	1.6	1.6	4
7	22	364	1.6	1.6	6	3.4	3.4	4
8	23	348	1.1	1.1	10	2.3	2.3	4
9	25	320	0.1	0.1	1	0.6	0.6	4
10	28	286	1	1	13	0.8	0.8	4
11	30	267	0.7	0.7	7	1	1	4
12	31	258	1.5	1.5	12	3.6	3.6	4
13	34	235	1.3	1.3	6	0.2	0.2	4
14	36	222	0.6	0.6	4	4	4	4
15	39	205	1.8	1.8	11	2.6	2.6	4
16	40	200	1.4	1.4	14	1.9	1.9	4
17	42	190	0.9	0.9	6	0.4	0.4	4
18	43	186	2	2	7	1.3	1.3	4
19	45	178	1.7	1.7	5	0.7	0.7	4
20	47	170	0.2	0.2	15	4.1	4.1	4
21	49	163	0.4	0.4	6	2.5	2.5	4
22	51	157	1.3	1.3	2	3.8	3.8	4
23	52	154	1.8	1.8	4	0.5	0.5	4
24	54	148	1.5	1.5	14	1.4	1.4	4
25	58	138	1	1	13	2.7	2.7	4
26	60	133	2	2	13	3.1	3.1	4
27	62	129	0.6	0.6	14	0.9	0.9	4
28	66	121	1.4	1.4	6	1.2	1.2	4
29	68	118	1.9	1.9	8	3.7	3.7	4
30	70	114	0.8	0.8	14	0.6	0.6	4

of 0.002 which is highlighted in bold, while the worst was experiment 12 with a generated error of 0.214 which is highlighted in italics.

In summary, the best architectures generated by the BSA algorithm are presented in Table 4.33.

Table 4.30 Architectures generated by the BSA for the systolic module

Systolic module					
No	Layers	Neurons		Epochs	AVG error
		Layer1	Layer2		
1	2	12	21	444	0.56
2	1	16		688	0.47
3	1	10		280	1.4
4	1	20		700	1.24
5	1	10		396	0.76
6	1	14		61	0.54
7	2	21	14	347	1.86
8	2	29	30	151	0.64
9	1	11		503	0.5
10	1	7		340	0.48
11	1	16		540	0.47
12	1	7		114	0.73
13	1	15		338	0.86
14	1	14		320	0.81
15	*2*	*30*	*1*	*183*	*6.53*
16	1	18		339	0.37
17	**1**	**12**		**346**	**0.49**
18	1	11		278	0.53
19	2	29	29	685	0.89
20	1	24		432	0.66
21	1	15		167	0.69
22	1	14		271	0.7
23	1	15		422	0.5
24	1	15		246	0.49
25	2	1	9	64	0.61
26	2	25	8	117	0.72
27	1	15		307	0.55
28	1	10		423	0.51
29	1	18		208	0.81
30	2	21	19	443	0.58

Table 4.31 Architectures generated by the BSA for the diastolic module

No	Layer	Neurons		Epochs	AVG error
		Layer1	Layer2		
1	2	2	13	522	0.003
2	1	4		687	0.004
3	2	15	17	111	0.004
4	1	19		205	0.023
5	1	19		361	0.007
6	1	25		117	0.006
7	1	29		477	0.004
8	*2*	*22*	*8*	*430*	*0.131*
9	2	7	3	482	0.004
10	1	28		677	0.003
11	2	30	16	488	0.003
12	2	21	2	205	0.004
13	2	9	20	700	0.003
14	2	13	18	469	0.012
15	2	13	17	495	0.01
16	2	25	8	185	0.003
17	2	8	6	448	0.004
18	1	11		645	0.003
19	2	29	2	417	0.004
20	1	7		438	0.006
21	1	4		625	0.003
22	1	28		283	0.004
23	1	24		238	0.009
24	2	4	13	566	0.004
25	1	14		268	0.004
26	2	23	1	522	0.003
27	1	19		287	0.003
28	2	30	11	252	0.004
29	**2**	**4**	**5**	**565**	**0.003**
30	2	7	22	270	0.004

Table 4.32 Architectures generated by the BSA for the heart rate module

Heart rate module

No	Layer	Neurons		Epochs	AVG error
		Layer1	Layer2		
1	1	22		665	0.003
2	1	25		637	0.003
3	1	7		364	0.004
4	1	10		144	0.004
5	2	13	27	688	0.005
6	2	6	4	493	0.003
7	1	13		175	0.003
8	2	30	17	422	0.003
9	1	27		138	0.003
10	2	28	27	180	0.003
11	2	8	20	472	0.004
12	2	24	9	540	0.214
13	1	8		611	0.003
14	2	17	5	335	0.203
15	1	17		229	0.003
16	1	17			0.008
17	2	24	21	221	0.003
18	2	13	14	513	0.004
19	**2**	**4**	**15**	**659**	**0.002**
20	1	6		352	0.004
21	2	14	22	510	0.246
22	2	26	23	464	0.003
23	1	21		52	0.003
24	2	27	3	505	0.004
25	2	28	16	464	0.003
26	2	8	10	577	0.008
27	2	10	17	297	0.003
28	2	18	29	148	0.004
29	1	8		483	0.015
30	2	28	26	552	0.015

Table 4.33 Summary of the best architectures generated by the BSA

Systolic		Diastolic		Heart rate	
Layers	1	Layers	2	Layers	2
Neurons per layers	12	Neurons per layers	4 5	Neurons per layers	4 15
Epochs	346	Epochs	565	Epochs	659
MSE	0.8133	MSE	0.1273	MSE	0.1883

4.3.2.2 Optimization of Modular Neural Network Using FPA

30 experiments are carried out to test the FPA algorithm, where parameter variation was performed in each experiment, which is presented in Table 4.34. The parameters that the algorithm uses are the following:

ind represents the number of individuals, *iteration* is the number of iterations, *Dim* is the number of dimensions, which is fixed since this is the parameter that moves the architecture of the neural network and *p* represents the switching probability.

The first experimentation carried out is the one corresponding to the module to obtain the trend of systolic pressure. Table 4.35 presents the variations of the architecture made by the FPA in the hidden layers and the number of neurons per each hidden layer, in the same way, the error obtained from performing each experiment is presented in the last column.

It can be analyzed that the best result was given in experiment 4 with an average error of 0.0001 which is highlighted in bold, while the worst experiment was number 21 with a mean error of 0.0023 and which is highlighted with italics.

The second experiment corresponds to the optimization of the module to obtain the trend of the diastolic pressure, in which variation is made in the parameters in each experiment, using those presented in Table 4.34. The results obtained are presented in Table 4.36, where it can be analyzed that the best experiment was 1 with an error of 3.79E-05 which is highlighted in bold, while the worst was experiment 21 with an error generated of 0.03701 which is highlighted in italics.

The third experiment carried out corresponds to the optimization of the module to obtain the heart rate trend, in which variation is made in the parameters in each experiment, using those presented in Table 4.34. The results obtained are presented in Table 4.37, where it can be analyzed that the best experiment was 1 with an error of 0.0002 which is highlighted in bold, while the worst was experiment 21 with a generated error of 0.002 which is highlighted in italics.

In summary, the best architectures generated by the FPA algorithm are presented in Table 4.38.

Table 4.34   Parameters used by the FPA in each experiment

Parameters

No	Ind	Iter	Dim	Prob
1	10	800	4	0.3
2	12	667	4	0.7
3	14	571	4	0.9
4	15	533	4	0.8
5	18	444	4	0.2
6	20	400	4	0.5
7	22	364	4	0.4
8	23	348	4	0.8
9	25	320	4	0.9
10	28	286	4	0.7
11	30	267	4	0.6
12	31	258	4	0.4
13	34	235	4	0.2
14	36	222	4	0.1
15	39	205	4	0.8
16	40	200	4	0.8
17	42	190	4	0.2
18	43	186	4	0.3
19	45	178	4	0.7
20	47	170	4	0.6
21	49	163	4	0.8
22	51	157	4	0.3
23	52	154	4	0.6
24	54	148	4	0.9
25	58	138	4	0.1
26	60	133	4	0.8
27	62	129	4	0.7
28	66	121	4	0.9
29	68	118	4	0.4
30	70	114	4	0.8

4.3.2.3   Comparative Analysis of the Results

Once the best architectures are obtained, the training that generates the best results is used to perform the simulation. 36 patients from the database to perform the test are taken, the readings obtained by the ABPM of these being the input to each module of the modular neural network. Experiments for the systolic pressure module with 15

Table 4.35 Architectures generated by the FPA for the systolic pressure module

Systolic module

No	Layer	Neurons		Epochs	AVG error
		Layer1	Layer2		
1	2	12	10	362	0.0002
2	2	17	1	450	0.0002
3	2	12	13	516	0.0003
4	**2**	**18**	**11**	**604**	**0.0001**
5	1	3		328	0.0002
6	2	30	26	98	0.0004
7	1	30		700	0.0009
8	2	23	24	288	0.0002
9	2	17	10	591	0.0007
10	1	7		1	0.0002
11	1	30		477	0.0005
12	2	19	30	700	0.0003
13	1	30		325	0.0004
14	2	26	30	700	0.0005
15	1	18		673	0.0007
16	2	9	20	234	0.0005
17	2	23	19	70	0.0007
18	2	11	3	387	0.0016
19	2	1	11	1	0.0011
20	1	7		5	0.0005
21	*1*	*25*		*700*	*0.0023*
22	2	17	1	124	0.0007
23	2	18	11	2	0.0009
24	1	24		663	0.0011
25	2	8	1	494	0.0005
26	1	30		365	0.0012
27	2	7	13	537	0.0009
28	2	12	28	150	0.0012
29	1	1		1	0.0007
30	1	4		1	0.0013

Table 4.36 Architectures generated by the FPA for the diastolic pressure module

No	Layers	Neurons		Epochs	AVG error
		Layer1	Layer2		
1	**2**	**30**	**6**	**118**	**0.0000379**
2	1	3		295	0.000034
3	1	7		658	0.0000513
4	1	30		20	0.0000537
5	2	5	26	120	0.0000834
6	2	18	4	465	0.000109
7	2	6	9	208	0.000064
8	2	10	6	700	0.000062
9	2	30	12	1	0.0000499
10	2	9	16	50	0.0000897
11	1	5		538	0.000118
12	1	7	21	363	0.000193
13	2	13	16	554	0.000108
14	1	28		115	0.00012
15	1	30		458	0.000127
16	1	19		700	0.0000823
17	2	3	25	455	0.000137
18	2	16	5	180	0.000134
19	1	19		405	0.000199
20	1	30		6	0.000178
21	*2*	*19*	*6*	*455*	*0.03701*
22	1	6		15	0.00017
23	1	21		328	0.000161
24	1	12		464	0.000168
25	1	30		471	0.000234
26	2	23	26	123	0.000218
27	1	16		284	0.000204
28	1	30		5	0.000284
29	1	30		693	0.000203
30	2	14	2	512	0.000204

Table 4.37 Architectures generated by the FPA for the heart rate module

Heart rate module

No	Layers	Neurons		Epochs	AVG Error
		Layer1	Layer2		
1	**1**	**21**		**693**	**0.0002**
2	1	30		587	0.0002
3	1	29		261	0.0005
4	1	8		689	0.0003
5	2	30	14	203	0.0006
6	2	30	30	1	0.0007
7	2	25	28	700	0.0007
8	1	30		288	0.0003
9	2	10	18	332	0.0004
10	2	22	15	437	0.001
11	2	30	30	315	0.0007
12	1	29		398	0.0007
13	1	24		530	0.0008
14	2	30	17	700	0.0012
15	2	23	13	255	0.0007
16	2	5	30	674	0.001
17	2	1	7	600	0.0012
18	2	1	9	252	0.001
19	1	30		71	0.0008
20	1	25		108	0.0012
21	*1*	*14*		*601*	*0.002*
22	1	30		639	0.0009
23	2	14	16	672	0.0013
24	2	1	9	700	0.0013
25	1	20		683	0.0015
26	2	8	22	409	0.0018
27	1	2		560	0.0015
28	1	25		15	0.0014
29	1	14		248	0.0015
30	2	19	12	354	0.0015

Table 4.38 Summary of the best architectures generated by the FPA

Systolic		Diastolic		Heart rate	
Layers	2	Layers	2	Layers	1
Neurons per layer	18 11	Neurons per layer	30 6	Neurons per layer	21
Epochs	604	Epochs	118	Epochs	693
MSE	1.267	MSE	0.446	MSE	0.602

patients are presented in Table 4.39, according to the expert, there can be a variation in the trend of 10 points, and as can be analyzed, good results are obtained in most cases to observe that less error is obtained with the BSA algorithm.

In Table 4.40 the experiments for the diastolic pressure modulus with 15 patients are presented. It is observed that good results are obtained in most cases, in addition to taking into account that less error is obtained with the BSA algorithm.

In Table 4.41 the experiments for the module to obtain the heart rate trend with 15 patients are presented. It is observed that good results are obtained in most cases, in addition to taking into account that less error is obtained with the BSA algorithm.

Table 4.42 summarize the percentage of success obtained by both algorithms; if the results are analyzed, it can be concluded that the BSA is better compared to the FPA algorithm for the 3 modules of the modular neural network.

Table 4.39 Simulation of the optimized MNN using 15 patients for the systolic module

No	Real	BSA			FPA		
		Trend	ABS E	% success	Trend	ABS E	% success
1	106	106	0	100	103	3	97
2	107	110	3	97	103	4	94
3	134	131	3	98	129	5	76
4	121	117	4	97	114	7	85
5	106	109	3	97	92	14	95
6	110	109	1	99	117	8	85
7	130	133	3	98	132	1	74
8	117	125	8	94	120	1	78
9	117	117	0	100	121	4	83
10	113	119	6	95	117	2	81
11	121	118	3	98	119	2	82
12	123	120	3	98	120	3	81
13	102	100	2	98	97	5	99
14	121	125	4	97	127	2	76
15	117	120	3	98	122	2	80

Table 4.40 Simulation of the optimized MNN using 15 patients for the diastolic module

No	Real	BSA			FPA		
		Trend	ABS E	% success	Trend	ABS E	% success
1	62	65	3	95	62	0	100
2	61	61	0	100	61	0	100
3	62	62	0	100	63	1	98
4	77	76	1	99	72	5	94
5	65	65	0	100	59	6	91
6	69	73	4	95	69	0	100
7	86	84	2	98	88	2	98
8	73	70	3	96	76	3	96
9	54	55	1	98	50	4	93
10	72	71	1	99	74	2	97
11	78	80	2	98	84	6	93
12	82	87	5	94	80	2	98
13	62	66	4	94	61	1	98
14	70	66	4	94	67	3	96
15	68	67	1	99	68	0	100

Table 4.41 Simulation of the optimized MNN using 15 patients for the heart rate module

No	Real	BSA			FPA		
		Trend	ABS E	% success	Trend	ABS E	% success
1	95	93	2	98	99	4	96
2	69	71	2	97	70	1	99
3	72	67	5	93	77	5	94
4	80	85	5	94	84	4	95
5	73	74	1	99	69	4	95
6	63	59	4	94	66	3	95
7	74	73	1	99	74	0	100
8	71	67	4	94	65	6	92
9	70	68	2	97	70	0	100
10	67	65	2	97	68	1	99
11	77	74	3	96	83	6	93
12	71	69	2	97	70	1	99
13	74	75	1	99	72	2	97
14	70	68	2	97	73	3	96
15	62	62	0	100	57	5	92

Table 4.42   Percent of success with the optimized MNN

Systolic		Diastolic		Heart rate	
BSA (%)	FPA (%)	BSA (%)	FPA (%)	BSA (%)	FPA (%)
96	84	**98**	96	**97**	95

Table 4.43   Percent of success with the non-optimized MNN

Systolic	Diastolic	Heart rate
94%	89%	95%

Table 4.43 presents the percentage of success obtained by the non-optimized modular neural network, where, if we compare the results obtained with the BSA, an improvement is observed.

4.4   Optimization of the Artificial Neural Network Used to Obtain the Risk of Developing Hypertension

For this fourth case study, the optimization of an artificial neural network is carried out to obtain the risk of developing hypertension in four years, using the following methodology.

4.4.1   Proposed Method for the Optimization of the Monolithic Neural Network Used to Obtain the Risk of Developing Hypertension

The proposed method presented in Fig. 4.32 has an artificial neural network, which is part of a neuro-fuzzy hybrid model described in [3–5]; it provides a medical diagnosis derived from the blood pressure of different patients. The neural network is inspired by the Framingham Heart Study, which offers patients the risk of developing hypertension in an incoming future period. Risk factors taken as inputs of the neural network are the following: systolic and diastolic pressure, age, gender, body mass index, if the patient is a smoker, and if he has hypertensive parents. With the information mentioned before the neural network is trained to learn the variation of the risk factors in each person to obtain as an output the percentage of risk that the person has in developing hypertension in 4 years.

To provide a more accurate diagnosis to patients using the neural network, the optimization of its architecture is carried out, and for this, the Flower Pollination Algorithm is used, and from which its performance is observed and make a comparison with the Simple Enumeration Method (SEM) to observe which one provides

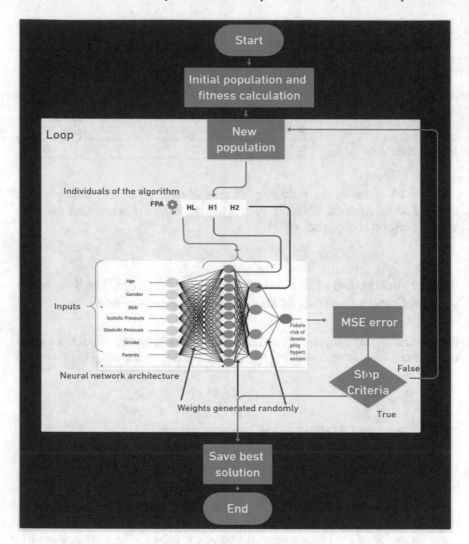

Fig. 4.32 Model used for optimization of the monolithic neural network

better results. The SEM has all the possible architecture combinations of neurons per hidden layer and the number of hidden layers.

Different parts of the model have been optimized using another bio-inspired algorithm [6–10]. In this case, the goal was to test the FPA algorithm, which makes the necessary adjustments in the architecture of the neural network, this means, that varying the number of hidden layers and the number of neurons in the hidden layers, for which a database with 500 patients is used to train the neural network in each experiment, this database has the risk factors mentioned before.

Neural Network Parameters			
H. Layers		Neurons H.L 1	Neurons H.L 2
1	2	1 - 30	1 – 30

Fig. 4.33 Neural network architecture representation in FPA

The individuals, which in the case of FPA will be the pollen to make the necessary changes in the architecture to obtain the one that provides the lower error when simulating the information.

As previously mentioned, the optimization of the architecture of a neural network was carried out, the FPA algorithm was used, which adjusted the parameters of the architecture, which were the number of hidden layers, due to the complexity of the problem was limited to 2. In addition to the number of neurons per hidden layer, which were limited to 30, this is due to the information with which we are working.

In Fig. 4.33, the representation of the individuals of the FPA algorithm is presented. The objective function used for the optimization was the MSE, to reduce the error in the training, which is found in Eq. 4.1.

4.4.2   Results Obtained from the Optimization

To analyze its operation and test the performance of the algorithm, 30 experiments are performed. Taking into account that the parameters of the algorithm are changed in each experiment to observe and analyze its performance and to obtain the best parameters for solve in an optimal way.

In Table 4.44 the variation of the parameters is presented, *ind* represents the number of individuals, *iteration* is the number of iterations, *Dim* is the number of dimensions, which is fixed, since this is the parameter that moves the architecture of the neural network and *p* represents the switch probability.

Table 4.45 presents the results obtained for each experiment, this means, the variation of the architecture made to the neural network by the algorithm, where the number of hidden layers and the number of neurons in each hidden layer were changed. In this case, experiment 1 was the one with the best-obtained result, which is highlighted in bold with an error of 5.78E-04, and experiment 19 being the worst with an error of 2.83E-03, and which is highlighted in italic.

Best-obtained architecture is summarized as follows:

- Number of layers: 2
- Neurons in hidden layer 1: 6
- Neurons in hidden layer 2: 4.

Figure 4.34 illustrates the comparison of the targets with the outputs generated by the neural network, with these 50 patients the neural network was tested in each

Table 4.44 FPA parameters used in each experiment

No.E	Individuals	Iteration	Dimensions	Probability
1	10	93	3	0.2
2	12	78	3	0.7
3	14	66	3	0.6
4	16	58	3	0.4
5	18	52	3	0.1
6	20	47	3	0.8
7	22	42	3	0.9
8	24	39	3	0.3
9	26	36	3	0.5
10	28	33	3	0.7
11	30	31	3	0.2
12	32	29	3	0.8
13	34	27	3	0.4
14	36	26	3	0.5
15	38	24	3	0.6
16	40	23	3	0.8
17	42	22	3	0.6
18	44	21	3	0.4
19	46	20	3	0.9
20	48	19	3	0.2
21	50	19	3	0.3
22	52	18	3	0.7
23	54	17	3	0.8
24	56	17	3	0.1
25	58	16	3	0.6
26	60	16	3	0.2
27	62	15	3	0.8
28	64	15	3	0.4
29	66	14	3	0.5
30	68	14	3	0.3

iteration of the algorithm, where the red line represents the target and the blue line the output generated by the neural network at the moment of simulating the information.

Once having the best architecture, the simulation is carried out with a group of 20 real patients. Table 4.46 presents the different risk factors of each one of them and where the comparison of the real results of the study is made, with the one obtained by the optimized neural network and the non-optimized neural network.

Table 4.45 FPA results

Exp	Layers	Neuros L1	Neuros L2	Epochs	Error	Time in minutes
1	**2**	**6**	**4**	**250**	**5.775E-04**	**17.76**
2	2	7	3	250	1.215E-03	20.78
3	2	15	3	250	1.379E-03	16.26
4	2	8	4	250	1.691E-03	18.50
5	2	9	3	250	1.007E-03	18.38
6	2	9	4	250	1.197E-03	17.83
7	2	4	29	250	1.580E-03	19.21
8	2	8	4	250	8.410E-04	20.69
9	2	8	3	250	1.168E-03	17.81
10	2	10	6	250	9.203E-04	18.14
11	2	10	3	250	1.886E-03	19.72
12	2	4	7	250	1.411E-03	18.12
13	2	6	5	250	8.606E-04	19.33
14	2	8	3	250	1.184E-03	16.78
15	2	9	5	250	8.228E-04	21.02
16	2	10	5	250	1.663E-03	21.03
17	2	8	6	250	1.751E-03	25.40
18	2	5	9	250	1.152E-03	21.32
19	2	6	8	250	2.833E-03	18.64
20	2	11	4	250	1.734E-03	20.80
21	2	9	3	250	1.069E-03	18.58
22	2	7	6	250	1.103E-03	20.71
23	2	8	5	250	1.321E-03	19.58
24	2	7	5	250	1.203E-03	23.01
25	2	10	4	250	1.384E-03	20.03
26	2	9	4	250	1.561E-03	21.99
27	2	5	5	250	1.725E-03	19.07
28	2	11	7	250	2.202E-03	23.58
29	2	8	4	250	1.422E-03	21.05
30	2	7	6	250	1.730E-03	22.14

Figure 4.35 illustrates the comparison of the real results with the output generated by the neural network for the 20 real patients, where the red line represents the target and the blue line the output generated by the neural network at the moment of simulating the information.

In Table 4.47, the statistical parameters of the FPA experiments are presented. In this, can be observed the value of the best and the worst experiments, the average

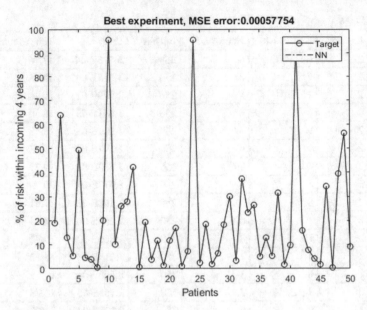

Fig. 4.34 Results of training with 50 patients

and standard deviation of all the 30 experiments that are taken into account for comparison purposes.

In Fig. 4.36, the convergence of the first ten experiments using the FPA algorithm are presented, it can be observed graphically that experiment 1 is the one where the least error was obtained, while that experiment 4 was the worst.

In Fig. 4.37, the second set of convergence experiments are illustrated, in this case, experiment 15 was the one that converged faster, but the error is not the best one obtained, while that experiment 19 is the worst of all-30 experiments.

Figure 4.38 presents the last set of experiments, for this, experiment 21 is the best, and the worst is experiment 28.

4.4.3 Z-test of FPA and ALO Versus Simple Enumeration Method

To compare and verify which of the different methods yields the least error, a statistical test is performed, using the Z test, which uses the formula 4.2.

In the first case, the null hypothesis states that the errors obtained by the FPA algorithm are greater than or equal to the errors obtained by the simple enumeration method (SEM). While the alternative hypothesis states that the errors obtained by the FPA algorithm are less than the errors obtained by the SEM. Table 4.48 illustrates

Table 4.46 Comparison results between non-optimized and optimized neural network

No	Age	Gen	BMI	Syst	Dias	Smoke	H.par	Real	Optimized nn	Non-optimized nn
1	27	M	24.30	122	77	No	0	4	4	4
2	28	M	23.36	120	81	No	0	5	5	5
3	28	W	29.76	123	82	No	2	16	16	18
4	25	M	24.40	114	65	No	0	1	1	1
5	45	M	24.90	116	75	No	1	6	6	6
6	31	W	35.26	95	61	No	1	0	0	0
7	33	M	25.26	130	74	No	2	10	10	11
8	32	M	29.98	123	76	No	1	7	7	8
9	25	M	21.70	108	66	No	0	0	0	1
10	30	M	30.30	123	78	No	0	7	7	7
11	30	W	21.55	107	61	No	1	1	1	1
12	32	M	24.49	112	72	No	0	2	2	2
13	31	W	30.07	112	71	No	2	3	3	3
14	29	W	21.50	99	62	No	0	0	0	1
15	31	W	23.40	106	65	No	0	1	1	2
16	26	W	31.90	126	68	No	0	4	4	4
17	32	M	31.10	110	68	No	0	1	1	1
18	30	M	28.91	122	76	No	0	5	5	5
19	31	W	29.00	114	66	No	2	2	2	3
20	27	M	22.72	115	72	No	1	2	2	2

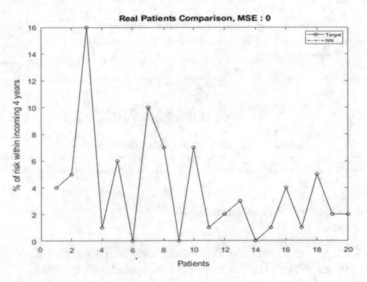

Fig. 4.35 Results from comparison with real patients

Table 4.47 Statistical parameters of experiments with FPA

Best	0.000577535
Worst	0.002833456
Average	0.001386461
Standard Deviation	0.000457444

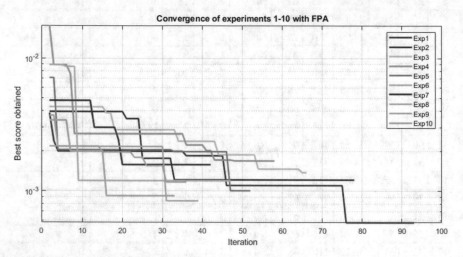

Fig. 4.36 FPA Convergence of experiments 1–10

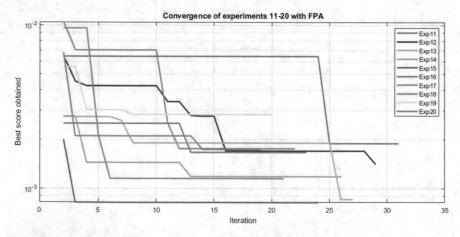

Fig. 4.37 FPA convergence of experiments 11–20

a comparison of means between the two methods with 30 experiments; it can be observed that the FPA obtained a lower error compared with the SEM.

Table 4.49 presents the statistical parameters for this test.

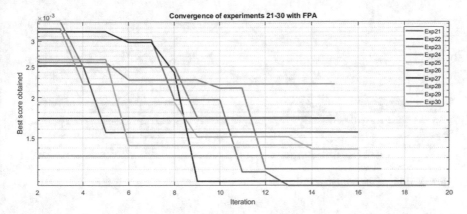

Fig. 4.38 FPA convergence of experiments 21–30

Table 4.48 Comparison results between methods

MSE	FPA	SEM
Average	0.001386461	0.0035578
Standard deviation	0.000457444	0.0015705
Experiments	30	30

Table 4.49 Statistical parameters for FPA vs SEM

Z test parameters FPA vs SEM	
Test statistic value	-7.251
Critical value (Z_c)	-1.64
Significance level (α)	0.05
H_0	$\mu 1 \geq \mu 2$
H_a (Claim)	$\mu 1 < \mu 2$
Experiments	30

Since it is observed that z test statistic value $z = -7.251$ is less than the critical value $z_c = -1.64$, it is then concluded that the null hypothesis is rejected and the alternative hypothesis is accepted. So, it can be concluded that there is sufficient evidence with a 5% level of significance to support the claim of the errors obtained by the FPA algorithm are lower than the errors obtained by the SEM; in Fig. 4.39 the probability distribution graph for this test is illustrated.

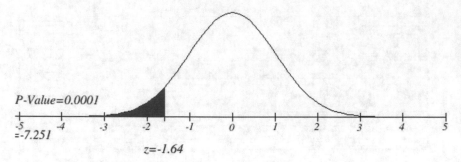

Fig. 4.39 Plot of the normal distribution

4.5   Optimization of the Modular Neural Network to Obtain the Risk of Developing a Cardiovascular Event

In the fifth case study, the optimization of a modular neural network is carried out to obtain the cardiovascular risk, of which the methodology carried out is explained as follows:

4.5.1   Proposed Method for Optimizing the Modular Network for the Risk of Developing a Cardiovascular Event

This is the last part that was added to the neuro-fuzzy model because the need arises to also know the risk that a patient has in developing a cardiovascular event in 10 years, since as has been observed in recent times this is one of the leading causes of death worldwide [11], and by being able to provide this information to the patient, it gives a diagnosis in time so that the persons can change them healthy habits. Another diagnosis that it can give in this part of the model is the age of the heart since the cardiologist needs to know it because with this it can be analyzed more clearly certain cardiovascular diseases that may occur. To provide the aforementioned diagnoses, a modular neural network is used, which consists of 3 modules, which are given as input different risk factors such as age, sex, systolic pressure, body mass index, if the patient is a smoker, if has diabetes and if is undergoing hypertensive treatment, each of these modules will learn the behavior of the information provided and as an output from the first module, the risk of developing hypertension is obtained in 10 years, the output of the second module corresponds to the age of the heart when the patient who consumes hypertensive treatment and the third module provides the age of the heart of the patient does not consume the hypertensive treatment, this modularity is carried out because it was difficult for the network to learn the behavior of the heart age if it provided information indiscriminately from patients who were using or not using hypertensive therapy, and it is necessary to provide a timely diagnosis.

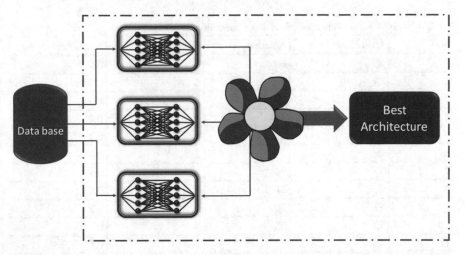

Fig. 4.40 Optimization of modular neural network for obtaining the risk of cardiovascular event

Due to the above, the modular neural network is optimized, to find the architecture that generates a better diagnosis. To do this, two bio-inspired algorithms are used, to observe its performance and analyze which one provides a better result. The algorithms used were the FPA and the BSA, which have been used in the different optimizations of the model with good results. The architecture parameters to be optimized are the number of hidden layers and the number of neurons in each hidden layer, due to previous experience, the search is limited to two hidden layers and 30 neurons per hidden layer, since, due to the nature of the information, with more than 30 neurons, begins to overtrain the modular neural network and provide unwanted results. Each module was trained with 250 patients in each iteration and was tested with 50 patients. In Fig. 4.40 the optimization of the modular neural network is graphically represented.

To find out which of the architectures provides a better result, the MSE which is found in Eq. 4.1.

4.5.2 Experimentation and Results of the Optimization

To determine which of the algorithms used gives us a better result, 30 experiments are carried out, where the parameters used by each one of them are varied, to determine with which of these variants the algorithm efficiently solves the problem provided.

Table 4.50 presents the parameters of the FPA algorithm, which are described in Sect. 4.3.

In the same way, 30 experiments are carried out with the BSA algorithm varying its parameters, which are presented in Table 4.51, and which have been described in Sect. 4.2.

Table 4.50 FPA parameters used in the experimentation

N.E	Individuals	Iteration	Dimension	Probability
1	15	62	3	0.9
2	17	55	3	0.3
3	19	49	3	0.5
4	21	44	3	0.7
5	23	40	3	0.3
6	25	37	3	0.2
7	27	34	3	0.8
8	29	32	3	0.6
9	31	30	3	0.9
10	33	28	3	0.1
11	35	27	3	0.4
12	37	25	3	0.8
13	39	24	3	0.2
14	41	23	3	0.7
15	43	22	3	0.9
16	45	21	3	0.5
17	47	20	3	0.3
18	49	19	3	0.2
19	51	18	3	0.5
20	53	18	3	0.9
21	55	17	3	0.3
22	57	16	3	0.6
23	59	16	3	0.8
24	61	15	3	0.7
25	63	15	3	0.9
26	65	14	3	0.4
27	67	14	3	0.3
28	69	13	3	0.9
29	71	13	3	0.5
30	73	13	3	0.7

The results obtained by the FPA algorithm are presented in Table 4.52, where the number of layers and neurons per layer provided by the algorithm can be analyzed, in the same way, the error generated when performing each experiment can be observed, which is obtained by testing the architectures obtained with 50 patients. For the first module, the best experiment was number 3, generating an error of 1.7622E-03, in the second module the best experiment was 4 with an error of 2.0687E-05 and for

Table 4.51 Parameters variation of BSA for experiments

No. E	Population	Iterations	a1	a2	Frequency	Cognitive A.C	Social A.C	Dimensions
1	16	58	0.2	0.2	3	0.4	0.4	3
2	18	52	1.33	1.33	8	0.6	0.6	3
3	20	47	0.9	0.9	1	1.4	1.4	3
4	22	42	0.4	0.4	10	1.8	1.8	3
5	24	39	0.1	0.1	12	0.2	0.2	3
6	26	36	1.2	1.2	7	1.9	1.9	3
7	28	33	1	1	6	2	2	3
8	30	31	1.42	1.42	2	3.48	3.48	3
9	32	29	0.75	0.75	10	1.22	1.22	3
10	34	27	1.25	1.25	9	3.2	3.2	3
11	36	26	1.95	1.95	15	3.15	3.15	3
12	38	24	0.6	0.6	5	0.8	0.8	3
13	40	23	1.3	1.3	12	2.84	2.84	3
14	42	22	0.9	0.9	14	0.5	0.5	3
15	44	21	1.6	1.6	7	1.44	1.44	3
16	46	20	1.4	1.4	5	4	4	3
17	48	19	0.9	0.9	15	0.9	0.9	3
18	50	19	0.6	0.6	11	1.18	1.18	3
19	52	18	2	2	14	2.4	2.4	3
20	54	17	1	1	10	3	3	3
21	56	17	0.4	0.4	11	0.9	0.9	3
22	58	16	0.9	0.9	11	2	2	3
23	60	16	1.6	1.6	5	1.88	1.88	3
24	62	15	0.4	0.4	6	3.12	3.12	3
25	64	15	1.2	1.2	7	2.5	2.5	3
26	66	14	1.8	1.8	2	3.44	3.44	3
27	68	14	0.7	0.7	5	3.7	3.7	3
28	70	13	0.5	0.5	11	2.83	2.83	3
29	72	13	2	2	6	1.56	1.56	3
30	74	13	1.7	1.7	14	4	4	3

the third module the best experiment was the 3 with an error obtained of 4.9969E-06. The best results are marked in bold, while the worst results are marked in italics.

Table 4.53 lists the best architectures found by the FPA algorithm for each of the modules. To complement the information on the training of the modules, it is necessary to mention that in each experiment 250 epochs were used and the Levenberg–Marquardt was used as the learning algorithm.

Table 4.52 Results obtained for each module using FPA

No	Module CR				Module HAT				Module HAW			
	Lyr	Neurons		Error	Lyr	Neurons		Error	Lyr	Neurons		Error
		L1	L2			L1	L2			L1	L2	
1	2	18	2	2.7139E-03	2	2	30	1.0560E-04	1	18		6.1907E-05
2	2	17	2	2.0452E-03	1	17		1.6883E-04	1	16		6.0907E-05
3	**2**	**13**	**2**	**1.7622E-03**	*1*	*12*		3.7648E-04	**2**	**2**	**18**	**4.9969E-06**
4	2	20	2	2.4270E-03	**2**	**3**	**22**	**2.0687E-05**	2	2	13	3.8364E-05
5	2	18	3	2.5873E-03	1	11		2.7427E-04	1	14		2.0338E-04
6	2	14	2	2.8492E-03	2	2	24	9.7008E-05	2	2	17	1.2372E-04
7	2	22	2	2.8170E-03	1	9		1.7437E-04	1	2		3.4650E-04
8	2	12	5	3.7180E-03	1	9		1.5743E-04	1	2		1.1126E-04
9	2	11	2	3.0731E-03	1	13		2.3490E-05	2	2	24	1.8725E-05
10	2	13	2	3.5421E-03	1	14		6.5299E-05	1	2		1.5807E-05
11	2	10	2	3.5495E-03	2	3	28	1.3000E-04	2	2	*11*	3.2050E-04
12	2	11	2	3.6995E-03	1	13		1.5528E-04	2	2	25	1.8010E-04
13	2	20	2	3.8815E-03	2	2	26	2.6078E-04	2	2	29	2.0071E-04
14	2	19	2	3.4877E-03	2	4	13	2.6881E-04	2	2	30	3.0541E-05
15	2	22	2	5.5936E-03	1	5		2.4263E-04	2	3	15	2.2829E-04
16	2	12	3	3.5020E-03	2	2	21	3.0814E-04	1	7		1.9435E-04
17	2	9	4	3.9976E-03	1	18		2.9908E-04	2	2	17	8.4739E-05
18	2	16	3	6.4363E-03	1	12		1.7657E-04	1	3		2.6109E-04
19	2	16	2	3.5655E-03	1	12		1.7728E-04	2	2	13	8.7971E-05

(continued)

Table 4.52 (continued)

No	Module CR				Module HAT				Module HAW			
	Lyr	Neurons		Error	Lyr	Neurons		Error	Lyr	Neurons		Error
		L1	L2			L1	L2			L1	L2	
20	2	15	2	6.8452E-03	1	10		2.8193E-04	2	2	28	9.8195E-06
21	2	15	3	1.8191E-03	1	22		1.0354E-04	1	2		5.8135E-05
22	2	12	3	4.5308E-03	2	2	16	8.0574E-05	2	2	14	8.1766E-05
23	2	14	2	3.2716E-03	2	2	25	2.4468E-04	1	2		3.1255E-05
24	2	21	2	2.2232E-03	2	2	30	9.0135E-05	1	2		1.1597E-05
25	2	15	2	3.8664E-03	1	3		2.6500E-04	1	4		1.4851E-05
26	2	9	5	5.8655E-03	2	2	21	1.1130E-04	1	4		6.5163E-05
27	2	13	5	4.5445E-03	2	2	26	2.6406E-04	2	2	10	2.0655E-05
28	2	12	2	3.4840E-03	1	21		2.6028E-04	1	7		2.2507E-04
29	2	14	2	2.8449E-03	2	5	30	2.0397E-04	1	3		9.1806E-05
30	2	20	2	2.7755E-03	1	2		3.0767E-04	2	2	27	1.5269E-05

Table 4.53 Summary of the best architectures

	Module CR	Module HAT	Module HAW
Layers number	2	2	2
Neurons per layer	13, 2	3,22	2,18

The results obtained from using the BSA algorithm for the optimization of each module are presented in Table 4.54. Where it can analyze the number of layers and numbers of neurons per layer thrown by each experiment performed. The error obtained in each experiment is also presented, as with the FPA algorithm, each generated architecture was tested with 50 patients. The best results are highlighted in bold, while the worst results are presented in italics.

The best architectures obtained by the BSA algorithm are summarized in Table 4.55. Similarly, in this experimentation 250 epochs were used and as a learning algorithm the Levenberg–Marquardt was used, since it has been proven that it generates excellent results when working with numerical data. In this case, for the first module, the best experiment was number 10, generating an error of 1.9151E-03, in the second module the best experiment was 15 with an error of 1.1581E-06 and for the third module the best experiment was the 20 with an error obtained of 1.3104E-06.

Once the experimentation or obtaining the best architectures is carried out, the best training is taken to be tested by 15 patients from their database. The results obtained are compared with the results provided by the Framingham Heart Study calculator because it takes it as a basis because it is well known that it is a reliable source of comparison. Table 4.56 presents the information on the risk factors of the group of patients studied.

The results obtained by the optimized modular neural network are presented in Table 4.57. The results generated by the Framingham Heart Study are available for comparison. In the same way, there are the results obtained from testing the best architectures provided by both algorithms, where it can be analyzed that the results obtained by the FPA algorithm are better than those obtained by the BSA algorithm where it is observed that more errors are generated.

Table 4.58 presents a comparison of the percentage of success obtained when testing the 15 patients from their database using the architectures optimized for each algorithm. With the information obtained, a comparison is made of the results generated with the optimized and non-optimized modular neural networks, where a significant improvement is observed when optimizing. And it can also determine that the architectures generated by the FPA algorithm provide a more accurate result than the results contained by the architecture generated by the BSA algorithm.

Table 4.54 Results obtained for each module using BSA

No	Module CR				Module HAT				Module HAW			
	Lyr	Neurons		Error	Lyr	Neurons		Error	Lyr	Neurons		Error
		L1	L2			L1	L2			L1	L2	
1	2	15	2	1.9895E-03	1	11		2.6692E-04	1	1		1.2594E-05
2	2	13	3	2.1560E-03	1	11		2.4896E-05	1	3		3.5618E-04
3	2	13	2	2.6370E-03	2	2	25	2.7216E-04	1	2		1.2674E-05
4	2	6	19	1.5795E-02	1	13		2.1723E-04	1	4		2.1344E-04
5	2	14	3	2.9891E-03	1	6		9.7679E-04	1	2		1.4953E-05
6	2	15	3	2.4868E-03	1	6		5.5475E-05	1	5		2.8574E-04
7	2	14	4	3.0008E-03	1	7		1.8742E-04	2	5	29	5.9159E-05
8	2	10	3	2.0662E-03	1	16		1.4719E-04	1	6		1.0244E-03
9	2	15	3	3.3236E-03	1	6		9.5376E-04	1	12		1.6154E-04
10	2	11	3	**1.9151E-03**	2	3	21	6.0404E-04	2	2	14	5.2981E-06
11	2	13	3	2.6533E-03	2	7	30	1.2470E-04	2	1	29	7.7887E-05
12	2	12	4	2.1487E-03	2	2	12	2.7296E-04	1	4		7.7950E-05
13	2	15	3	3.0417E-03	1	6		8.9198E-05	1	8		2.2912E-05
14	2	7	9	5.8475E-03	1	5		1.8147E-04	1	11		2.4948E-04
15	2	14	4	3.4193E-03	**1**	**11**		**1.1581E-06**	1	4		1.3707E-05
16	2	14	3	2.9070E-03	1	18		5.0817E-05	1	4		3.7762E-05
17	2	12	4	3.4450E-03	1	9		6.9434E-05	1	3		1.2314E-04
18	2	9	3	1.2373E-02	1	13		9.2995E-05	1	2		2.8704E-05
19	2	13	4	2.5051E-03	1	15		3.4264E-04	1	26		4.2738E-04

(continued)

Table 4.54 (continued)

No	Module CR				Module HAT				Module HAW			
	Lyr	Neurons		Error	Lyr	Neurons		Error	Lyr	Neurons		Error
		L1	L2			L1	L2			L1	L2	
20	2	12	3	4.5717E-03	1	5		1.0276E-04	**2**	**2**	**27**	**1.3104E-06**
21	2	8	9	7.0193E-03	1	11		5.0799E-04	1	3		5.1014E-06
22	2	14	3	3.1124E-03	1	11		1.2354E-04	2	2	30	4.8564E-05
23	2	19	3	3.7331E-03	2	8	29	4.9013E-05	2	1	9	2.5942E-05
24	2	18	3	4.1533E-03	1	9		5.5214E-05	2	2	25	3.5750E-05
25	2	12	2	2.3808E-03	1	6		1.1892E-04	1	1		1.9581E-05
26	2	19	2	1.9537E-03	2	2	5	4.0356E-04	2	1	28	1.0944E-04
27	2	10	5	2.1283E-03	1	14		3.2272E-04	1	1		1.5986E-05
28	2	9	4	4.3478E-03	1	8		1.0511E-04	1	5		4.3936E-05
29	2	15	3	2.3804E-03	1	8		6.1481E-05	1	8		2.2459E-05
30	2	20	1	4.0480E-03	2	4	30	1.7215E-04	2	1	30	3.6029E-05

Table 4.55 Results obtained for each module using BSA

No	Module CR				Module HAT				Module HAW			
	Lyr	Neurons		Error	Lyr	Neurons		Error	Lyr	Neurons		Error
		L1	L2			L1	L2			L1	L2	
1	2	15	2	1.9895E-03	1	11		2.6692E-04	1	1		1.2594E-05
2	2	13	3	2.1560E-03	1	11		2.4896E-05	1	3		3.5618E-04
3	2	13	2	2.6370E-03	2	2	25	2.7216E-04	1	2		1.2674E-05
4	2	6	19	1.5795E-02	1	13		2.1723E-04	1	4		2.1344E-04
5	2	14	3	2.9891E-03	1	6		9.7679E-04	1	2		1.4953E-05
6	2	15	3	2.4868E-03	1	6		5.5475E-05	1	5		2.8574E-04
7	2	14	4	3.0008E-03	1	7		1.8742E-04	2	5	29	5.9159E-05
8	2	10	3	2.0662E-03	1	16		1.4719E-04	1	6		1.0244E-03
9	2	15	3	3.3236E-03	1	6		9.5376E-04	1	12		1.6154E-04
10	2	**11**	3	**1.9151E-03**	2	3	21	6.0404E-04	2	2	14	5.2981E-06
11	2	13	3	2.6533E-03	2	7	30	1.2470E-04	2	1	29	7.7887E-05
12	2	12	4	2.1487E-03	2	2	12	2.7296E-04	1	4		7.7950E-05
13	2	15	3	3.0417E-03	1	6		8.9198E-05	1	8		2.2912E-05
14	2	7	9	5.8475E-03	1	5		1.8147E-04	1	11		2.4948E-04
15	2	14	4	3.4193E-03	**1**	**11**		**1.1581E-06**	1	4		1.3707E-05
16	2	14	3	2.9070E-03	1	18		5.0817E-05	1	4		3.7762E-05
17	2	12	4	3.4450E-03	1	9		6.9434E-05	1	3		1.2314E-04
18	2	9	3	1.2373E-02	1	13		9.2995E-05	1	2		2.8704E-05
19	2	13	4	2.5051E-03	1	15		3.4264E-04	1	26		4.2738E-04

(continued)

Table 4.55 (continued)

No	Module CR				Module HAT				Module HAW			
	Lyr	Neurons		Error	Lyr	Neurons		Error	Lyr	Neurons		Error
		L1	L2			L1	L2			L1	L2	
20	2	12	3	4.5717E-03	1	5		1.0276E-04	**2**	**2**	**27**	**1.3104E-06**
21	2	8	9	7.0193E-03	1	11		5.0799E-04	1	3		5.1014E-06
22	2	14	3	3.1124E-03	1	11		1.2354E-04	2	2	30	4.8564E-05
23	2	19	3	3.7331E-03	2	8	29	4.9013E-05	2	1	9	2.5942E-05
24	2	18	3	4.1533E-03	1	9		5.5214E-05	2	2	25	3.5750E-05
25	2	12	2	2.3808E-03	1	6		1.1892E-04	1	1		1.9581E-05
26	2	19	2	1.9537E-03	2	2	5	4.0356E-04	2	1	28	1.0944E-04
27	2	10	5	2.1283E-03	1	14		3.2272E-04	1	1		1.5986E-05
28	2	9	4	4.3478E-03	1	8		1.0511E-04	1	5		4.3936E-05
29	2	15	3	2.3804E-03	1	8		6.1481E-05	1	8		2.2459E-05
30	2	20	1	4.0480E-03	2	4	30	1.7215E-04	2	1	30	3.6029E-05

Table 4.56 List of risk factors in a group of patients

Patient	Risk factors						
	Age	Gender	Body Mass index	Systolic BP	Diabetes?	Smoke?	Treatment
1	27	Woman	24.3	122	No	No	No
2	28	Woman	23.36	120	No	No	No
3	28	Man	29.76	123	No	No	No
4	25	Woman	24.4	114	No	No	No
5	45	Woman	24.9	116	No	No	No
6	31	Man	35.26	95	No	No	No
7	33	Man	25.26	130	No	No	No
8	32	Woman	29.98	123	No	No	No
9	25	Woman	21.7	108	No	No	No
10	30	Woman	30.3	123	No	No	No
11	30	Man	21.55	107	No	No	No
12	32	Woman	24.49	112	No	No	No
13	31	Man	30.07	112	No	No	No
14	29	Man	21.5	99	No	No	No
15	31	Man	23.4	106	No	No	No

Table 4.57 Results obtained for the best architectures

Patient	FHS		FPA			BSA		
	Risk percentage	Age of heart	Risk percentage	Age of heart	ABS value	Risk percentage	Age of heart	ABS value
1	0.8	27	0.8	27	0	0.8	27	0
2	0.8	28	0.8	28	0	0.8	28	0
3	1.7	28	1.7	28	0	1.8	28	0.1
4	0.5	25	0.5	25	0	0.5	25	0
5	2.8	45	2.8	45	0	2.8	45	0
6	1.7	31	1.7	31	0	1.7	31	0
7	1.7	33	1.7	33	0	1.6	33	0.1
8	1.4	32	1.4	32	0	1.4	32	0
9	0.4	25	0.4	25	0	0.4	25	0
10	1.2	30	1.2	30	0	1.2	30	0
11	1.3	30	1.3	30	0	1.3	30	0
12	1	32	1	32	0	1	32	0
13	2	31	2	31	0	2.1	31	0.1
14	1	29	1	29	0	1.1	29	0.1
15	1.5	31	1.5	31	0	1.5	31	0

Table 4.58 Success percent of the experiments

Non-optimized Modular NN	Optimized modular NN	
	FPA	BSA
85%	100%	96%

Table 4.59 Parameters used with BSA and FBSA

	Iterations	Population	Dimensions	Frequency	Cognitive A.C	Social A.C	a1	a2
BSA	1000	30	20	3	1.5	1.5	1	1
FBSA	1000	30	20	3	Dynamic	Dynamic	1	1

4.6 Fuzzy Bird Swarm Algorithm

In this last case study, a dynamic parameter adjustment is carried out on the BSA algorithm to improve its performance, because in different experiments this was the algorithm that did not provide the desired results, and for this, 3 study cases are carried out. The proposed method and the performed experiments are described below.

4.6.1 Proposed Method for the Dynamic Parameter Adjustment

A fuzzy dynamic parameters adjustment is provided to the BSA algorithm, which will be given the name Fuzzy Bird Swarm Algorithm because when this algorithm is compared with the FPA algorithm, it did not provide good results in some study cases. Derived from doing different experiments with the algorithm parameters, it is determined to adjust the $c1$ and $c2$ parameters, because a significant change is observed when these parameters vary, which correspond to the cognitive and social acceleration coefficients, respectively. The proposed method with the dynamic adjustment of parameters is presented in Fig. 4.41.

Three Mamdani-type fuzzy systems are designed with trapezoidal membership functions to perform the proposed method, in which the variation that is made is in the rules. The scheme of the fuzzy system is presented in Fig. 4.42.

The input corresponds to the iterations, for this, the percentage of the current iteration with respect to the total iterations is calculated, to better understand this point, when the algorithm starts to execute the iterations take a low value and as increase, this takes a higher value. This behavior is presented in Eq. 4.3 and implemented initially by [12].

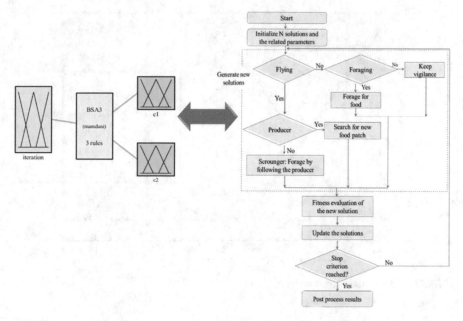

Fig. 4.41 Proposed method for the dynamic parameter adjustment

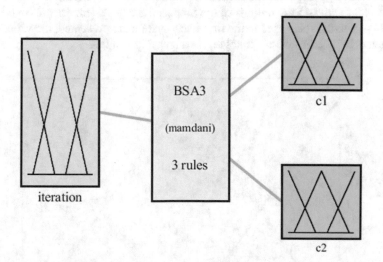

Fig. 4.42 Fuzzy system for the dynamic parameter adjustment

$$iteration = \frac{Current\ iteration}{Total\ nummber\ iterations} \qquad (4.3)$$

The iteration input is granulated in 3, using the linguistic variables "Low", "Medium" and "High". The image of the input variable is presented in Fig. 4.43.

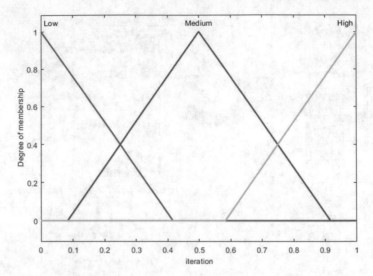

Fig. 4.43 Input variable Iteration

Regarding the outputs, which correspond to the parameters $c1$ and $c2$, these are granulated in 3, using both the linguistic variables "Low", "Medium" and "High". In Figs. 4.44 and 4.45 the outputs of the proposed fuzzy system are presented.

In Fig. 4.46 the rules used in the first fuzzy system are presented, the value of the parameters to be adjusted are changing in increasing fashion.

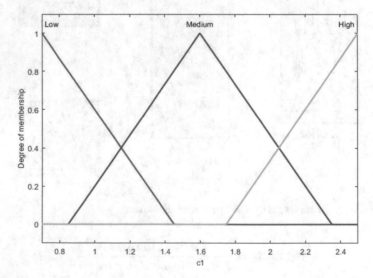

Fig. 4.44 Output variable c1

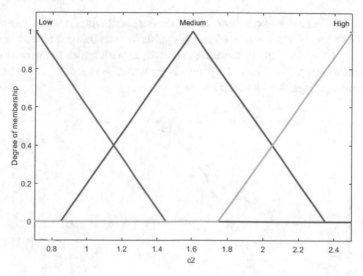

Fig. 4.45 Output variable c2

1. If *iteration* is *Low* then *c1* is *Low* and *c2* is *Low*

2. If *iteration* is *Medium* then *c1* is *Medium* and *c2* is *Medium*

3. If *iteration* is *High* then *c1* is *High* and *c2* is *High*

Fig. 4.46 Rules of fuzzy system 1

The rules used in the second fuzzy system, where c1 is increasing and c2 is decreasing, are presented in Fig. 4.47.

In Fig. 4.48 the rules used by the fuzzy system are presented where *c1* is decreasing and *c2* is increasing.

1. If *iteration* is *Low* then *c1* is *Low* and *c2* is *High*

2. If *iteration* is *Medium* then *c1* is *Medium* and *c2* is *Medium*

3. If *iteration* is *High* then *c1* is *High* and *c2* is *Low*

Fig. 4.47 Rules of fuzzy system 2

1. If *iteration* is *Low* then *c1* is *High* and *c2* is *Low*

2. If *iteration* is *Medium* then *c1* is *Medium* and *c2* is *Medium*

3. If *iteration* is *High* then *c1* is *Low* and *c2* is *High*

Fig. 4.48 Rules of fuzzy system 3

Similarly, a Mamdani-type fuzzy system is designed with Gaussian membership functions, where the same parameters are used as the one designed with trapezoidal membership functions, this to compare and analyze with which the performance of the algorithm is improved. The fuzzy input is presented in Fig. 4.49 while the outputs are presented in Figs. 4.50 and 4.51.

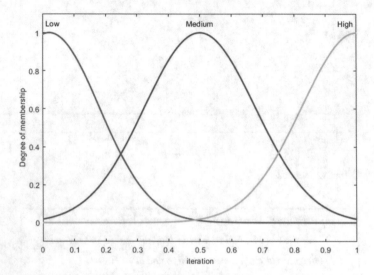

Fig. 4.49 Input variable iteration with Gaussian MFs

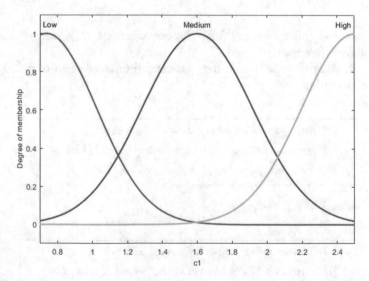

Fig. 4.50 Output variable c1 with Gaussian MFs

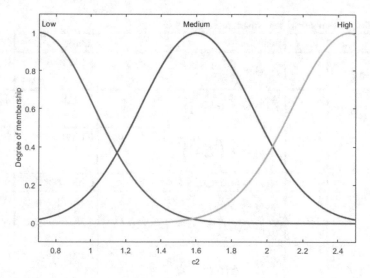

Fig. 4.51 Output variable c2 with Gaussian MFs

4.6.2 Experiments and Results

Experiments are carried out with 3 case studies to observe with which fuzzy system the best result is obtained. For the first study, experiments are carried out with 10 mathematical functions, of which 7 are unimodal and 3 are multimodal. Table 4.59 presents the parameters used to carry out the corresponding experimentation, in which the first row corresponds to the BSA algorithm and the second row to the proposed method, these parameters of BSA were described in Sect. 4.2.

Table 4.60 presents the set of mathematical functions used in this first experiment. Column 1 describes the type of function, column 3 shows the mathematical function. The ranges and minimum values of these functions are presented in columns 4 and 5, respectively, where it is observed that in all cases the minimum is 0.

In the second case study, experiments are carried out with 10 functions corresponding to CEC2017 competition, for this 100 independent runs were carried out for each function, in addition to performing 10 experiments per function. The parameters used are presented in Table 4.61.

In Table 4.62, the different mathematical functions used for experimentation are presented, for this, column 1 contains the type of function with which are working, the second column corresponds to the number of the function, column 3 contains the names of the functions used, and finally, column 4 contains *Fi*, which corresponds to the minimum value of each function. It is worth mentioning that the ranges of these functions go from −100 to 100.

For the third case study, the proposed method is applied to the optimization of an artificial neural network, which provides the diagnosis of the risk of developing

Table 4.60 Mathematical benchmark problems

	No	Function	Range	$f$min
Unimodal Benchmark functions	1	$f_1(x) = \sum\limits_{i=1}^{n} x_i^2$	[-100,100]	0
	2	$f_2(x) = \sum\limits_{i=1}^{n} \|x_i\| + \prod\limits_{i=1}^{n} \|x_i\|$	[-10,10]	0
	3	$f_3(x) = \sum\limits_{i=1}^{n} \left(\sum\limits_{j-1}^{i} x_j \right)^2$	[-100,100]	0
	4	$f_4(x) = max_i\{\|x_i\|, , 1 \leq i \leq n\}$	[-100,100]	0
	5	$f_5(x) = \sum\limits_{i=1}^{n-1} \left[100\left(x_{i+1} - x_i^2\right)^2 + (x_i - 1)^2 \right]$	[-30,30]	0
	6	$f_6(x) = \sum\limits_{i=1}^{n} ([x_i + 0.5])^2$	[-100,100]	0
	7	$f_7(x) = \sum\limits_{i=1}^{n} ix_i^4 + random[0, 1]$	[-1.28, 1.28]	0
Multimodal benchmark functions	8	$f_9(x) = \sum\limits_{i=1}^{n} \left[x_i^2 - 10cos(2\pi x_i) + 10 \right]$	[-5.12,5.12]	0
	9	$f_{10}(x) = -20exp\left(-0.2\sqrt{\frac{1}{n} \sum\limits_{i=1}^{n} x_i^2}\right) - exp\left(\frac{1}{n} \sum\limits_{i=1}^{n} cos(2\pi x_i)\right) + 20 + e$	[-32,32]	0
	10	$f_{11}(x) = \frac{1}{400} \sum\limits_{i=1}^{n} x_i^2 - \prod\limits_{i=1}^{n} cos\left(\frac{x_i}{\sqrt{i}}\right) + 1$	[-600,600]	0

Table 4.61 Parameters used in second case of study

	Iterations	Population	Dimensions	Frequency	Cognitive A.C	Social A.C	a1	a2
BSA	1500	30	30	3	1.5	1.5	1	1
FBSA	1500	30	30	3	Dynamic	Dynamc	1	1

hypertension in a time and from which the number of layers and the number of neurons in each layer.

This artificial neural network is based on a neuro-fuzzy hybrid model [3, 13], which provides different results to provide a final medical diagnosis. From the database generated and which was already explained in Chap. 3, information is taken from the patients, such as gender, body mass index, age, systolic and diastolic blood pressure, if the patient has hypertensive parents and if the patient is a smoker, these will serve as an entrance to said artificial neural network. Once having this,

Table 4.62 CEC2017 functions

	No	Function	Fi
Unimodal Benchmark functions	5	Shifted and Rotated Rastrigin's Function	500
	6	Shifted and Rotated Expanded Scaffer's F6 Function	600
	7	Shifted and Rotated Lunacek Bi Rastrigin Function	700
	8	Shifted and Rotated Non-Continuous Rastrigin's Function	800
	9	Shifted and Rotated Levy Function	900
	10	Shifted and Rotated Schwefel's Function	1000
Hybrid benchmark functions	11	Hybrid Function 1 (N = 3)	1100
Multimodal benchmark functions	21	Composition Function 1 (N = 3)	2100
	22	Composition Function 2 (N = 3)	2200
	23	Composition Function 3 (N = 4)	2300

$[-100, 100]$

the FBSA algorithm looks for the parameters where the network generates the least error at the time of providing the results, this means, with which combination of parameters the network best learns the behavior of the information. Having previous experience in the optimization of different neural networks for diagnosis and trend, it is limited to 2 hidden layers and 30 neurons for each neuron, because it has been proven that by increasing the number of these parameters the neural network is over trained providing unwanted results. The representation of the optimization of the artificial neural network is presented in Fig. 4.52. As in previous optimizations, the MSE is used as the objective function.

4.6.3 Results

The average of the results corresponding to the first case study is presented in Table 4.63. The information is organized as follows: Column 1 determines the number of the function used for the experimentation, the second column corresponds to the results obtained by the original method, columns 3 to 8 presents the results obtained by the different proposed fuzzy systems. Analyzing the obtained information, it can be observed that, there is an improvement with respect to the original algorithm, but we still cannot determine with which of the fuzzy systems a better result is obtained since these are very similar.

Table 4.64 presents the averages of the results obtained in the experimentation of the second case study. In this, a significant improvement is also observed with respect

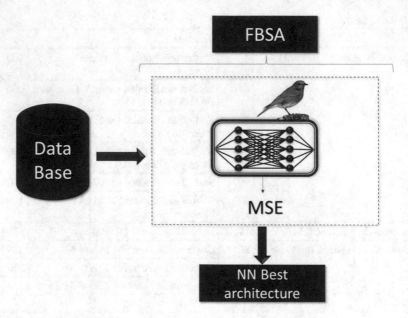

Fig. 4.52 Neural network optimization

Table 4.63 First case study results

No	Original algorithm	FBSA 1		FBSA 2		FBSA 3	
		Triangular	Gaussian	Triangular	Gaussian	Triangular	Gaussian
1	0	0	0	0	0	0	0
2	7.9E-212	8.8E-236	2.7E-241	5.2E-245	**4.2E-247**	2E-244	6E-246
3	0	0	0	0	0	0	0
4	1.4E-212	8.8E-237	3.4E-240	4.5E-243	1.1E-242	8.5E-245	**1.7E-246**
5	18.93802	8.616109	7.920785	**7.26079**	7.960495	7.544581	7.799847
6	3.457556	0.057171	0.041854	0.033016	0.025625	0.022677	**0.021388**
7	7.74E-05	7.38E-05	7.6E-05	7.6E-05	**7.3E-05**	7.68E-05	7.52E-05
9	0	0	0	0	0	0	0
10	0	0	0	0	0	0	0
11	0	0	0	0	0	0	0

to the results provided by the original method, and in the case of the FBSA, it is already clearly observed that the adjustment made with the fuzzy system number 3 provides better results. If only the results provided with the fuzzy system 3 are observed, it can be analyzed that with the fuzzy system with trapezoidal membership functions, good results are obtained in 5 mathematical functions, while with the fuzzy system that is designed with Gaussian membership functions, it provides better results. results in 3 mathematical functions.

Table 4.64 Second case study results

No	Original algorithm	FBSA 1		FBSA 2		FBSA 3	
		Triangular	Gaussian	Triangular	Gaussian	Triangular	Gaussian
5	8.396E+02	7.288E+02	7.401E+02	7.288E+02	7.285E+02	7.160E+02	**7.137E+02**
6	6.732E+02	6.862E+02	6.486E+02	6.438E+02	6.437E+02	**6.406E+02**	6.415E+02
7	1.355E+03	1.067E+03	1.097E+03	1.067E+03	1.063E+03	**1.045E+03**	1.049E+03
8	1.075E+03	9.935E+02	1.000E+03	9.935E+02	9.933E+02	9.917E+02	**9.857E+02**
9	7.602E+03	3.432E+03	4.091E+03	3.432E+03	3.445E+03	**3.296E+03**	3.337E+03
10	7.243E+03	7.138E+03	7.082E+03	7.138E+03	7.151E+03	7.190E+03	7.168E+03
11	5.349E+03	1.632E+03	1.767E+03	1.632E+03	1.639E+03	1.625E+03	**1.610E+03**
21	2.645E+03	2.503E+03	2.520E+03	2.503E+03	2.504E+03	**2.492E+03**	2.500E+03
22	8.184E+03	3.691E+03	3.804E+03	3.691E+03	3.670E+03	3.831E+03	3.716E+03
23	3.352E+03	2.997E+03	3.040E+03	2.997E+03	2.988E+03	**2.961E+03**	2.991E+03

For the last case study, 30 experiments are carried out, varying in each one of them the parameters of the algorithm, which are presented in Table 4.65 and which correspond to the original method.

The parameters used in the FBSA algorithm are presented in Table 4.66, where it is observed that $c1$ and $c2$ are dynamic.

The errors resulting from each experiment are presented in Table 4.67, where, when analyzing the results, it can be determined that the smallest error is provided by the FBSA, and in the same way, as with the set of functions of the CEC 2017, the best result is obtained with fuzzy system number 3 with trapezoidal membership functions.

Table 4.68 presents the averages of the errors obtained in the experimentation, and where it is observed that the best average of the error derived from the 30 experiments is provided by the fuzzy system number 1 with Gaussian membership functions.

The best obtained neural network architecture is presented in Table 4.69.

Table 4.65 Parameters used for BSA in the third case of study

N.E	Iterations	Population	Frequency	Cognitive A.C	Social A.C	a1	a2	Dimensions
1	400	10	11	0.5	0.5	2	2	3
2	333	12	5	0.8	0.8	1.5	1.5	3
3	285	14	14	1.2	1.2	0.4	0.4	3
4	250	16	11	1.5	1.5	0.1	0.1	3
5	222	18	6	1.8	1.8	0.8	0.8	3
6	200	20	4	2	2	1	1	3
7	181	22	14	2.33	2.33	1.3	1.3	3
8	166	24	15	2.48	2.48	0.6	0.6	3
9	153	26	6	2.76	2.76	0.9	0.9	3
10	142	28	4	3	3	1.1	1.1	3
11	133	30	9	3.18	3.18	1.9	1.9	3
12	125	32	14	3.22	3.22	0.5	0.5	3
13	117	34	4	3.45	3.45	1.5	1.5	3
14	111	36	10	3.56	3.56	0.7	0.7	3
15	105	38	7	4	4	1.3	1.3	3
16	100	40	10	0.4	0.4	1.8	1.8	3
17	95	42	3	0.7	0.7	0.3	0.3	3
18	90	44	11	1.15	1.15	0.9	0.9	3
19	86	46	7	1.34	1.34	1	1	3
20	83	48	3	1.45	1.45	2	2	3
21	80	50	5	1.67	1.67	0.6	0.6	3
22	76	52	7	1.78	1.78	0.3	0.3	3
23	74	54	4	1.92	1.92	1.5	1.5	3
24	71	56	5	2.18	2.18	1.2	1.2	3
25	68	58	8	2.39	2.39	1.8	1.8	3
26	66	60	14	2.56	2.56	0.7	0.7	3
27	64	62	4	2.83	2.83	0.9	0.9	3
28	62	64	8	3.4	3.4	1.5	1.5	3
29	58	68	11	3.7	3.7	1.7	1.7	3
30	57	70	7	4	4	2	2	3

4.6.4   Statistical Test

To prove that the results obtained by the FBSA are better than the BSA, a parametric statistical test is performed, which corresponds to the Z test, and is represented mathematically in Eq. 4.2.

Table 4.66 Parameter used by FBSA in the three-case study

N.E	Iterations	Population	Frequency	Cognitive A.C	Social A.C	a1	a2	Dimensions
1	93	10	11	Dynamic	Dynamic	2	2	3
2	78	12	5	Dynamic	Dynamic	1.5	1.5	3
3	66	14	14	Dynamic	Dynamic	0.4	0.4	3
4	58	16	11	Dynamic	Dynamic	0.1	0.1	3
5	52	18	6	Dynamic	Dynamic	0.8	0.8	3
6	47	20	4	Dynamic	Dynamic	1	1	3
7	42	22	14	Dynamic	Dynamic	1.3	1.3	3
8	39	24	15	Dynamic	Dynamic	0.6	0.6	3
9	36	26	6	Dynamic	Dynamic	0.9	0.9	3
10	33	28	4	Dynamic	Dynamic	1.1	1.1	3
11	31	30	9	Dynamic	Dynamic	1.9	1.9	3
12	29	32	14	Dynamic	Dynamic	0.5	0.5	3
13	27	34	4	Dynamic	Dynamic	1.5	1.5	3
14	26	36	10	Dynamic	Dynamic	0.7	0.7	3
15	24	38	7	Dynamic	Dynamic	1.3	1.3	3
16	23	40	10	Dynamic	Dynamic	1.8	1.8	3
17	22	42	3	Dynamic	Dynamic	0.3	0.3	3
18	21	44	11	Dynamic	Dynamic	0.9	0.9	3
19	20	46	7	Dynamic	Dynamic	1	1	3
20	19	48	3	Dynamic	Dynamic	2	2	3
21	19	50	5	Dynamic	Dynamic	0.6	0.6	3
22	18	52	7	Dynamic	Dynamic	0.3	0.3	3
23	17	54	4	Dynamic	Dynamic	1.5	1.5	3
24	17	56	5	Dynamic	Dynamic	1.2	1.2	3
25	16	58	8	Dynamic	Dynamic	1.8	1.8	3
26	16	60	14	Dynamic	Dynamic	0.7	0.7	3
27	15	62	4	Dynamic	Dynamic	0.9	0.9	3
28	15	64	8	Dynamic	Dynamic	1.5	1.5	3
29	14	66	11	Dynamic	Dynamic	1.7	1.7	3
30	14	68	7	Dynamic	Dynamic	2	2	3

For the experiments carried out with the functions of the CEC2017, the null hypothesis determines that the results obtained by the FBSA algorithm are greater than or equal to the results obtained with the BSA. As an alternative hypothesis, the results obtained by the FBSA are better than the results provided by the BSA. The statistical parameters of the second case study are presented in Table 4.70.

Table 4.67 Errors obtained in the optimization of the neural network

Exp	Original algorithm	FBSAv1		FBSA 2		FBSA3	
		Trapezoidal	Gaussian	Trapezoidal	Gaussian	Trapezoidal	Gaussian
1	5.7900E-04	2.3640E-04	1.1994E-04	1.2774E-04	2.1868E-04	1.1097E-04	1.1603E-04
2	*2.1700E-01*	1.5152E-04	1.4671E-04	**9.7513E-05**	5.9168E-04	2.9474E-04	**1.1203E-04**
3	6.9200E-04	3.6554E-04	1.6271E-04	1.0628E-04	1.6373E-04	1.3355E-04	1.1637E-04
4	6.6800E-04	1.8996E-04	1.3298E-04	1.8295E-04	2.0475E-04	4.1062E-04	1.4204E-04
5	5.8100E-04	2.0056E-04	1.0542E-04	1.1237E-04	4.5377E-04	1.3328E-04	1.1362E-04
6	1.2100E-03	*6.0916E-04*	2.0166E-04	1.5014E-04	2.7373E-04	1.7478E-04	1.8582E-04
7	5.2700E-04	2.0586E-04	1.2743E-04	1.6825E-04	1.2336E-04	1.6152E-04	1.6128E-04
8	7.4400E-04	1.4262E-04	2.2285E-04	1.6526E-04	1.3185E-04	1.3763E-04	1.5623E-04
9	7.9600E-04	2.7321E-04	1.1584E-04	1.7291E-04	1.5438E-04	1.1038E-04	1.5243E-04
10	7.0300E-04	1.2285E-04	1.9806E-04	1.7173E-04	1.3342E-04	7.5037E-05	1.1322E-04
11	5.1700E-04	1.8118E-04	1.3675E-04	1.6738E-04	1.7483E-04	7.7470E-04	**1.1203E-04**
12	5.8200E-04	5.8996E-04	8.4936E-05	1.7818E-04	1.1907E-04	1.1098E-04	1.5724E-04
13	9.1100E-04	1.4007E-04	1.2367E-04	*3.0270E-04*	1.9567E-04	1.2425E-04	2.6117E-04
14	8.4000E-04	1.2649E-04	2.1311E-04	1.3062E-04	1.5719E-04	**7.4491E-05**	2.0479E-04
15	4.6100E-04	1.2904E-04	1.3982E-04	1.3896E-04	2.2734E-04	1.7992E-04	1.1300E-04
16	4.3400E-04	2.7673E-04	1.7857E-04	1.6118E-04	1.7180E-04	1.6894E-04	1.4219E-04
17	6.1900E-04	2.2692E-04	1.4242E-04	1.3873E-04	1.1648E-04	2.3428E-04	1.5647E-04
18	7.1300E-04	1.3800E-04	1.5711E-04	2.2345E-04	1.6546E-04	1.0498E-04	1.7784E-04
19	**3.3800E-04**	1.2584E-04	1.6403E-04	1.7263E-04	1.8858E-04	1.8562E-04	2.1170E-04
20	8.7200E-04	3.3963E-04	1.4398E-04	1.8697E-04	2.2736E-04	1.7733E-04	1.4367E-04
21	1.1200E-02	1.5953E-04	1.6386E-04	2.6993E-04	1.8575E-04	*9.8745E-04*	1.7532E-04
22	5.8600E-04	**1.1478E-04**	*2.5263E-04*	1.0480E-04	1.5168E-04	1.6172E-04	2.0189E-04
23	5.9200E-04	1.2788E-04	1.2530E-04	1.1244E-04	1.1014E-04	1.3037E-04	1.1726E-04
24	7.0400E-04	1.6354E-04	1.9270E-04	2.4045E-04	4.1225E-04	1.4246E-04	*2.6490E-04*
25	6.5700E-03	2.1305E-04	1.2973E-04	1.5502E-04	1.5748E-04	1.5556E-04	1.4902E-04
26	5.6200E-04	1.8905E-04	1.5780E-04	2.0893E-04	1.2280E-04	1.0250E-04	1.6072E-04
27	6.5300E-04	1.4522E-04	1.7640E-04	1.8523E-04	1.4011E-04	9.6802E-05	2.2420E-04
28	7.6300E-04	1.7327E-04	**7.0441E-05**	1.5279E-04	2.3381E-04	1.4626E-04	1.3111E-04
29	7.4500E-04	1.2422E-04	2.0196E-04	1.7152E-04	1.5606E-04	1.5658E-04	1.5658E-04
30	7.2300E-04	3.0350E-04	1.7573E-04	1.4433E-04	2.7212E-04	2.0311E-04	2.0311E-04

Table 4.68 Average of the neural network optimization

Original	FBSA 1		FBSA 2		FBSA 3	
	Trapezoidal	Gaussian	Trapezoidal	Gaussian	Trapezoidal	Gaussian
8.43E-03	2.16E-04	**1.55E-04**	1.67E-04	2.05E-04	2.05E-04	1.61E-04

Table 4.69 Best Neural Network architecture

	Monolithic NN
Number of layers	2
Number of neurons per layer	8,6

Table 4.70 Statistical parameters for FBSA versus BSA

Z test parameters FBSA versus BSA	
Critical Value (Z_c)	−1.64
Significance Level (α)	0.05
H_0	$\mu1 \geq \mu2$
H_a (Claim)	$\mu1 < \mu2$
Level of significance	95%

The results obtained derived from applying the Z test are presented in Table 4.71, where the first column determines the lists the functions used, in columns 2 and 3 the average and standard deviation of the experiments carried out with the BSA are presented, in Columns 4 and 5 presents the averages and standard deviations of the experiments carried out by the FBSA, taking into account that it took the fuzzy system number 3 to perform this statistic. In column 5 the Z values are presented and finally, in column 6 the evidence of the obtained z values is determined, where "S" corresponds to the fact that significant evidence was obtained, while "NS" corresponds to that there was no significant evidence when performing the statistical test. Analyzing the information, it can be observed that in 9 of the 10 functions it finds significant evidence, so it can be concluded that the results obtained by the FBSA algorithm are better than those obtained by the BSA.

Table 4.71 Results of statistical test for CEC2017 functions

No	Original		FBSA		Z value	Evidence
	Average	Standard deviation	Average	Standard deviation		
5	839.55	4.30	712.74	7.44	−80.83	Significant
6	673.23	1.19	641.39	1.26	−100.75	Significant
7	1354.97	8.70	1044.43	8.41	−140.57	Significant
8	1075.02	6.52	986.26	3.16	−67.10	Significant
9	7601.72	217.87	3300.65	105.02	−97.40	Significant
10	7242.97	78.28	7240.19	119.18	−0.11	Not Significant
11	5348.57	335.48	1702.34	250.48	−47.70	Significant
21	2644.52	8.46	2486.35	4.51	−90.37	Significant
22	8183.92	96.92	3922.18	261.25	−83.77	Significant
23	3352.21	13.05	2980.71	13.70	−107.54	Significant

In the case of experimentation with the artificial neural network, it is established as a null hypothesis that the errors obtained by the FBSA algorithm are greater than the results obtained with the BSA algorithm. And as an alternative hypothesis, we have that the errors obtained by the FBSA are less than the errors contained by the BSA. The comparison of the averages obtained from both experiments is presented in Table 4.72, in which it can be noted that the errors of the FBSA are less than the errors generated by the FBSA.

The statistical parameters for this test are presented in Table 4.73.

Since a Z value of -11.472 is obtained and it is less than the critical value $Z_c = -1.64$, the null hypothesis is rejected and the alternative hypothesis is accepted, with which it is concluded that there is sufficient evidence with 5% level of significance to support the claim that the errors held by the FBSA are better than the errors obtained by the BSA. The probabilistic distribution of this case study is illustrated in Fig. 4.53.

Table 4.72 Comparison results between methods

MSE	FBSA	BSA
Average	1.55E-04	8.43E-03
Standard deviation	4.10E-05	3.95E-03
Experiments	30	30

Table 4.73 Statistical parameters for FBSA versus BSA

Z test parameters FBSA versus BSA	
Test Statistic Value	-11.472
Critical Value (Z_c)	-1.64
Significance Level (α)	0.05
H_0	$\mu 1 \geq \mu 2$
H_a (Claim)	$\mu 1 < \mu 2$
Experiments	30

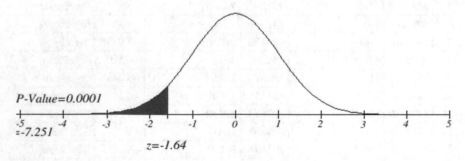

Fig. 4.53 Plot of the normal distribution

References

1. M.D. Feria-carot, J. Sobrino, Nocturnal hypertension. Hipertens. y riesgo Cardiovasc **28**(4), 143–148 (2011)
2. V. Marchione, Healthy resting heart rate by age for men and women. Bel Marra Health. 2018. https://www.belmarrahealth.com/resting-heart-rate-chart-factors-influence-heart-rate-elderly/. Accessed 12 Aug 2018
3. P. Melin, I. Miramontes, G. Prado-Arechiga, A hybrid model based on modular neural networks and fuzzy systems for classification of blood pressure and hypertension risk diagnosis. Expert Syst. Appl. **107**, 2018
4. P. Melin, J.C. Guzmán, G. Prado-Arechiga, Introduction to neuro fuzzy hybrid model, in *Neuro Fuzzy Hybrid Models for Classification in Medical Diagnosis*, ed. by P. Melin, J.C. Guzmán, G. Prado-Arechiga (Springer International Publishing, Cham, 2021), p. 103
5. I. Miramontes, G. Martínez, P. Melin, G. Prado-Arechiga, A hybrid intelligent system model for hypertension risk diagnosis, in *Fuzzy Logic in Intelligent System Design* (2018), pp. 202–213
6. I. Miramontes, P. Melin, G. Prado-Arechiga, Fuzzy system for classification of nocturnal blood pressure profile and its optimization with the crow search algorithm, in *Soft Computing Applications* (2021), pp. 23–34
7. P. Melin, G. Prado-Arechiga, I. Miramontes, O. Carvajal, Optimization of an artificial neuronal network for diagnosis of develop hypertension with 7 risk factors. J. Hypertens. **37**, 2019. https://journals.lww.com/jhypertension/Fulltext/2019/07001/OPTIMIZATION_OF_AN_ARTIFICIAL_NEURONAL_NETWORK_FOR.8.aspx
8. J.C. Guzmán, I. Miramontes, P. Melin, G. Prado-Arechiga, Optimal genetic design of type-1 and interval type-2 fuzzy systems for blood pressure level classification. Axioms **8**(1), 1–35 (2019)
9. I. Miramontes, P. Melin, G. Prado-Arechiga, Particle swarm optimization of modular neural networks for obtaining the trend of blood pressure, in *Intuitionistic and Type-2 Fuzzy Logic Enhancements in Neural and Optimization Algorithms: Theory and Applications*, ed. by O. Castillo, P. Melin, J. Kacprzyk (Springer International Publishing, Cham, 2020), pp. 225–236
10. I. Miramontes, J.C. Guzman, P. Melin, G. Prado-Arechiga, Optimal design of interval type-2 fuzzy heart rate level classification systems using the bird swarm algorithm. Algorithms **11**(12), 2018
11. World Health Organization (2017). https://www.who.int/news-room/fact-sheets/detail/cardiovascular-diseases-(cvds). Accessed 03 Dec 2020
12. F. Olivas, F. Valdez, O. Castillo, P. Melin, Dynamic parameter adaptation in particle swarm optimization using interval type-2 fuzzy logic. Soft Comput. **20**(3), 1057–1070 (2016)
13. I. Miramontes, G. Martínez, P. Melin, G. Prado-Arechiga, A hybrid intelligent system model for hypertension diagnosis, in *Nature-Inspired Design of Hybrid Intelligent Systems*, ed. by P. Melin, O. Castillo, J. Kacprzyk (Springer International Publishing, Cham, 2017), pp. 541–550

Chapter 5
Conclusions of the Hybrid Medical Model

For this book, a hybrid neuro-fuzzy model is proposed to provide a diagnosis based on blood pressure. Soft computing has proven to be an excellent tool to solve these types of problems, the deep learning carried out by artificial neural networks and the decisions made by fuzzy systems have been efficient when providing the diagnosis of a disease, such as COVID-19, Parkinson, diabetes, cancer, dengue, etc.

The proposed model can be identified in several parts in which artificial neural networks and fuzzy inference systems are used, which determine different conditions that a patient may have, determined by their blood pressure to result in a final medical diagnosis, and where each of these modules were optimized to provide an accurate diagnosis.

The first optimized technique was the fuzzy system, which provide us with information about the patient's heart rate level and the nocturnal blood pressure profile. These fuzzy systems were implemented in type-1 and type-2 fuzzy logic. Likewise, these were optimized by bio-inspired algorithms, where it is concluded that there was an improvement in the results obtained with the optimized fuzzy ones, achieving a 100% success rate with a specific group of patients.

Subsequently, the optimization of the different implemented artificial neural networks was carried out. First, the modular neural network was optimized to obtain the blood pressure trend, where in the same way it is optimized with two bio-inspired algorithms to observe its performance, where the results also improve significantly compared to the non-optimized modular neural network. This being the same case for the monolithic neural network with which the risk of developing hypertension is obtained in 4 years.

In another part of the model, an artificial neural network is implemented to determine the risk of a cardiovascular event in 10 years, with which different tests were carried out and it is determined that the artificial neural network learns better by dividing the information, this means to implement it in a modular way so that learning was better, in addition to being each module optimized and where results were improved.

From the algorithms implemented, it can be concluded that those with which the best results were obtained in the different optimizations carried out were the Flower Pollination Algorithm and the Bird Swarm Algorithm, which is why they were used throughout the investigation.

Finally, and derived from the fact that on different occasions the FPA algorithm provided better results, an improvement is made in the BSA algorithm by implementing dynamic adjustment of parameters with fuzzy systems where tests are performed with different mathematical functions and in the optimization of an artificial neural network, determining that the results obtained and the statistics implemented significantly improve the results is obtained.

In summary, the soft computing techniques implemented in the medical area have been extremely important and efficient in helping expert doctors make decisions to provide an accurate and timely diagnosis, in this case of hypertension and cardiovascular risk.

As future work, the database of patients can be expanded to have more different behaviors from blood pressure, also, at the time of giving the diagnosis, different recommendations given by the medical expert could be included, the designed fuzzy system can be tested with the General Type-2 fuzzy systems and finally, it would be to implement in hardware, some portable device so that cardiologists can transport it easily.

Regarding the dynamic adjustment of parameters to the BSA algorithm, diversity can be included as input of the fuzzy system to observe if the results improve, in addition to testing with the IT2FS to compare results.

Appendix A
Knowledge Representation

In this section, it is presented the representation of knowledge of Type-1 and Interval Type-2 fuzzy systems for heart rate classifiers of the nocturnal blood pressure profile, besides it can also find a brief description of the operation of the graphical user interface.

The following is the knowledge representation of the different fuzzy inference systems that yielded better results.

Type-1 Fuzzy System Knowledge Representation for Heart Rate Classification

In this section, the knowledge representation of Type-1 fuzzy system for obtaining the heart rate level with trapezoidal membership functions and optimized by BSA is presented.

Inputs Variables

The input Age uses *"Child"*, *"Young"*, *"Adult"* and *"Elder"* as linguistic variable. The values of each point of the fuzzy system are presented as follows.

© The Author(s), under exclusive license to Springer Nature Switzerland AG 2022
P. Melin et al., *Nature-inspired Optimization of Type-2 Fuzzy Neural Hybrid Models for Classification in Medical Diagnosis*, SpringerBriefs in Computational Intelligence, https://doi.org/10.1007/978-3-030-82219-4

$$\mu child(x) = \begin{cases} 0, x \leq 0 \\ 1, 0 \leq x \leq 5.684 \\ \frac{13.76-x}{8.076}, 5.684 \leq x \leq 13.76 \\ 0, 13.76 \leq x \end{cases} \quad (A.1)$$

$$\mu young(x) = \begin{cases} 0, x \leq 9 \\ \frac{x-9}{6.85}, 9 \leq x \leq 15.85 \\ 1, 15.85 \leq x \leq 31.12 \\ \frac{33.85-x}{2.73}, 31.12 \leq x \leq 33.85 \\ 0, 33.85 \leq x \end{cases} \quad (A.2)$$

$$\mu adult(x) = \begin{cases} 0, x \leq 32.15 \\ \frac{x-32.15}{11.6}, 32.15 \leq x \leq 43.75 \\ 1, 43.75 \leq x \leq 55.35 \\ \frac{60.91-x}{5.56}, 55.35 \leq x \leq 60.91 \\ 0, 60.91 \leq x \end{cases} \quad (A.3)$$

$$\mu elder(x) = \begin{cases} 0, x \leq 54.18 \\ \frac{x-54.18}{31.1}, 54.18 \leq x \leq 85.28 \\ 1, 85.28 \leq x \leq 100 \\ 0, 100 \leq x \end{cases} \quad (A.4)$$

The input Heart Rate variable uses *"Very Low"*, *"Low"*, *"Normal"*, *"High"* and *"VeryHigh"* as linguistic variable. The values of each point of the fuzzy system are given below:

$$\mu verylow(x) = \begin{cases} 0, x \leq 0 \\ 1, 0 \leq x \leq 12.38 \\ \frac{32.45-x}{19.07}, 13.38 \leq x \leq 32.45 \\ 0, 32.45 \leq x \end{cases} \quad (A.5)$$

$$\mu low(x) = \begin{cases} 0, x \leq 26.32 \\ \frac{x-26.32}{7.38}, 26.32 \leq x \leq 33.7 \\ 1, 33.7 \leq x \leq 48.51 \\ \frac{62.5-x}{13.99}, 48.51 \leq x \leq 62.5 \\ 0, 62.5 \leq x \end{cases} \quad (A.6)$$

$$\mu normal(x) = \begin{cases} 0, x \leq 58.63 \\ \frac{x-58.63}{10.13}, 58.63 \leq x \leq 68.76 \\ 1, 68.76 \leq x \leq 85.12 \\ \frac{98.71-x}{13.59}, 85.12 \leq x \leq 98.71 \\ 0, 98.71 \leq x \end{cases} \quad (A.7)$$

$$\mu high(x) = \begin{cases} 0, x \leq 90.12 \\ \frac{x-90.12}{23.58}, 90.12 \leq x \leq 113.7 \\ 1, 113.7 \leq x \leq 143.2 \\ \frac{158.9-x}{15.7}, 143.2 \leq x \leq 158.9 \\ 0, 158.9 \leq x \end{cases} \tag{A.8}$$

$$\mu veryhigh(x) = \begin{cases} 0, x \leq 143.5 \\ \frac{x-143.5}{50.1}, 143.5 \leq x \leq 193.6 \\ 1, 193.6 \leq x \leq 220 \\ 0, 220 \leq x \end{cases} \tag{A.9}$$

Output Variable

The output Heart Rate Level uses "*Low*", "*BeloWAV*", "*Excelent*", "*AboveAV*" and "*VeryHigh*" as linguistic variables. The values of each point of the fuzzy system are presented as follows.

$$\mu low(x) = \begin{cases} 0, x \leq 0 \\ 1, 0 \leq x \leq 4.977 \\ \frac{13.4-x}{8.423}, 4.977 \leq x \leq 13.4 \\ 0, 13.4 \leq x \end{cases} \tag{A.10}$$

$$\mu belowav(x) = \begin{cases} 0, x \leq 10.97 \\ \frac{x-10.97}{5.77}, 10.97 \leq x \leq 16.74 \\ 1, 16.74 \leq x \leq 23.96 \\ \frac{27.17-x}{3.21}, 23.96 \leq x \leq 27.17 \\ 0, 27.17 \leq x \end{cases} \tag{A.11}$$

$$\mu excellent(x) = \begin{cases} 0, x \leq 26.55 \\ \frac{x-26.55}{4.46}, 26.55 \leq x \leq 31.01 \\ 1, 31.07 \leq x \leq 39.06 \\ \frac{43.84-x}{4.78}, 39.06 \leq x \leq 43.84 \\ 0, 43.84 \leq x \end{cases} \tag{A.12}$$

$$\mu aboveav(x) = \begin{cases} 0, x \leq 39.42 \\ \frac{x-39.42}{8.54}, 39.42 \leq x \leq 47.96 \\ 1, 47.96 \leq x \leq 62.98 \\ \frac{71.81-x}{8.83}, 62.98 \leq x \leq 71.81 \\ 0, 71.81 \leq x \end{cases} \tag{A.13}$$

$$\mu veryhigh(x) = \begin{cases} 0, x \le 66.87 \\ \frac{x-66.87}{20.57}, 66.87 \le x \le 87.44 \\ 1, 87.44 \le x \le 100 \\ 0, 100 \le x \end{cases} \quad \text{(A.14)}$$

IT2FS Knowledge Representation Using Gaussian Membership Functions

The knowledge representation of the optimized IT2FS for obtaining the classification of the heart rate level using Gaussian membership functions is described in this section.

Inputs Variables

The input Age uses "*Child*", "*Young*", "*Adult*" and "*Elder*" as linguistic variables for each upper and lower membership functions. The values of each point of the fuzzy system are presented as follows.

$$\underline{\mu}child(x) = \alpha\exp\left[-\frac{1}{2}\left(\frac{x-0}{7.72}\right)^2\right] \quad \text{(A.15)}$$

$$\overline{\mu}child(x) = \exp\left[-\frac{1}{2}\left(\frac{x-0}{7.72}\right)^2\right] \quad \text{(A.16)}$$

$$\underline{\mu}young(x) = \alpha\exp\left[-\frac{1}{2}\left(\frac{x-23.54}{7.72}\right)^2\right] \quad \text{(A.17)}$$

$$\overline{\mu}young(x) = \exp\left[-\frac{1}{2}\left(\frac{x-23.54}{7.72}\right)^2\right] \quad \text{(A.18)}$$

$$\mu adult(x) = \alpha\exp\left[-\frac{1}{2}\left(\frac{x-46.13}{5.426}\right)^2\right] \quad \text{(A.19)}$$

$$\underline{\mu}adult(x) = \alpha\exp\left[-\frac{1}{2}\left(\frac{x-46.13}{5.426}\right)^2\right] \quad \text{(A.20)}$$

$$\underline{\mu}elder(x) = \alpha\exp\left[-\frac{1}{2}\left(\frac{x-100}{24.82}\right)^2\right] \quad \text{(A.21)}$$

$$\overline{\mu}elder(x) = \exp\left[-\frac{1}{2}\left(\frac{x-100}{24.82}\right)^2\right] \tag{A.22}$$

The Input Heart Rate variable uses "*Very Low*", "*Low*", "*Normal*", "*High*" and "*VeryHigh*" as linguistic variables for each upper and lower membership functions. The values of each point of the fuzzy system are presented as follows.

$$\underline{\mu}verylow(x) = \alpha\exp\left[-\frac{1}{2}\left(\frac{x-0}{10.91}\right)^2\right] \tag{A.23}$$

$$\underline{\mu}verylow(x) = \alpha\exp\left[-\frac{1}{2}\left(\frac{x-0}{10.91}\right)^2\right] \tag{A.24}$$

$$\underline{\mu}verylow(x) = \alpha\exp\left[-\frac{1}{2}\left(\frac{x-0}{10.91}\right)^2\right] \tag{A.25}$$

$$\overline{\mu}low(x) = \exp\left[-\frac{1}{2}\left(\frac{x-42.25}{9.739}\right)^2\right] \tag{A.26}$$

$$\underline{\mu}normal(x) = \alpha\exp\left[-\frac{1}{2}\left(\frac{x-77.94}{9.381}\right)^2\right] \tag{A.27}$$

$$\overline{\mu}normal(x) = \exp\left[-\frac{1}{2}\left(\frac{x-77.94}{9.381}\right)^2\right] \tag{A.28}$$

$$\underline{\mu}high(x) = \alpha\exp\left[-\frac{1}{2}\left(\frac{x-127.5}{18.94}\right)^2\right] \tag{A.29}$$

$$\underline{\mu}high(x) = \alpha\exp\left[-\frac{1}{2}\left(\frac{x-127.5}{18.94}\right)^2\right] \tag{A.30}$$

$$\underline{\mu}veryhigh(x) = \alpha\exp\left[-\frac{1}{2}\left(\frac{x-220}{40.86}\right)^2\right] \tag{A.31}$$

$$\overline{\mu}veryhigh(x) = \exp\left[-\frac{1}{2}\left(\frac{x-220}{40.86}\right)^2\right] \tag{A.32}$$

Output Variable

The output Heart Rate Level uses *"Low"*, *"BeloWAV"*, *"Excelent"*, *"AboveAV"* and *"VeryHigh"* as linguistic variables for each lower and upper membership functions. The values of each point of the fuzzy system are presented as follows.

$$\underline{\mu}low(x) = \alpha\exp\left[-\frac{1}{2}\left(\frac{x-0}{4.432}\right)^2\right] \tag{A.33}$$

$$\overline{\mu}low(x) = \exp\left[-\frac{1}{2}\left(\frac{x-0}{4.432}\right)^2\right] \tag{A.34}$$

$$\underline{\mu}belowav(x) = \alpha\exp\left[-\frac{1}{2}\left(\frac{x-19.55}{4.236}\right)^2\right] \tag{A.35}$$

$$\overline{\mu}belowav(x) = \exp\left[-\frac{1}{2}\left(\frac{x-19.55}{4.236}\right)^2\right] \tag{A.36}$$

$$\underline{\mu}excellent(x) = \alpha\exp\left[-\frac{1}{2}\left(\frac{x-35.38}{4.649}\right)^2\right] \tag{A.37}$$

$$\overline{\mu}excellent(x) = \exp\left[-\frac{1}{2}\left(\frac{x-35.38}{4.649}\right)^2\right] \tag{A.38}$$

$$\underline{\mu}aboveav(x) = \alpha\exp\left[-\frac{1}{2}\left(\frac{x-60.25}{7.076}\right)^2\right] \tag{A.39}$$

$$\overline{\mu}abovav(x) = \exp\left[-\frac{1}{2}\left(\frac{x-60.25}{7.076}\right)^2\right] \tag{A.39}$$

$$\underline{\mu}aboveav(x) = \alpha\exp\left[-\frac{1}{2}\left(\frac{x-100}{17.09}\right)^2\right] \tag{A.40}$$

$$\overline{\mu}abovav(x) = \exp\left[-\frac{1}{2}\left(\frac{x-100}{17.09}\right)^2\right] \tag{A.41}$$

Knowledge Representation of the Fuzzy Classifier to Obtain the Nocturnal Blood Pressure Profile

This part it describes the knowledge representation of the optimized Type-1 fuzzy system for the classification of the nocturnal blood pressure profile with trapezoidal membership functions.

Inputs Variables

The Input Systolic variable uses *"Low"*, *"Normal"*, *"High"*, and *"VeryHigh"* as linguistic variables. The values of each point of the fuzzy system are presented as follows.

$$\mu low(x) = \begin{cases} 0, x \leq 0.4 \\ 1, 0.4 \leq x \leq 0.6655 \\ \frac{0.8-x}{0.1345}, 0.6655 \leq x \leq 0.8 \\ 0, 0.8 \leq x \end{cases} \tag{A.42}$$

$$\mu normal(x) = \begin{cases} 0, x \leq 0.787 \\ \frac{x-0.787}{0.024}, 0.787 \leq x \leq 0.811 \\ 1, 0.811 \leq x \leq 0.889 \\ \frac{0.9102-x}{0.0212}, 0.889 \leq x \leq 0.9102 \\ 0, 0.9102 \leq x \end{cases} \tag{A.43}$$

$$\mu high(x) = \begin{cases} 0, x \leq 0.898 \\ \frac{x-0.898}{0.025}, 0.898 \leq x \leq 0.923 \\ 1, 0.923 \leq x \leq 0.9821 \\ \frac{1.02-x}{0.0379}, 0.9821 \leq x \leq 1.02 \\ 0, 1.02 \leq x \end{cases} \tag{A.44}$$

$$\mu veryhigh(x) = \begin{cases} 0, x \leq 1.001 \\ \frac{x-1.001}{0.089}, 1.001 \leq x \leq 1.09 \\ 1, 1.09 \leq x \leq 1.3 \\ 0, 1.3 \leq x \end{cases} \tag{A.45}$$

The Input Diastolic variable uses *"Low"*, *"Normal"*, *"High"*, and *"VeryHigh"* as linguistic variables. The values of each point of the fuzzy system are presented as follows.

$$\mu low(x) = \begin{cases} 0, x \leq 0.4 \\ 1, 0.4 \leq x \leq 0.6655 \\ \frac{0.8-x}{0.1345}, 0.6655 \leq x \leq 0.8 \\ 0, 0.8 \leq x \end{cases} \tag{A.46}$$

$$\mu normal(x) = \begin{cases} 0, x \leq 0.787 \\ \frac{x-0.787}{0.024}, 0.787 \leq x \leq 0.811 \\ 1, 0.811 \leq x \leq 0.889 \\ \frac{0.9102-x}{0.0212}, 0.889 \leq x \leq 0.9102 \\ 0, 0.9102 \leq x \end{cases} \quad (A.47)$$

$$\mu high(x) = \begin{cases} 0, x \leq 0.898 \\ \frac{x-0.898}{0.025}, 0.898 \leq x \leq 0.923 \\ 1, 0.923 \leq x \leq 0.9821 \\ \frac{1.02-x}{0.0379}, 0.9821 \leq x \leq 1.02 \\ 0, 1.02 \leq x \end{cases} \quad (A.48)$$

$$\mu veryhigh(x) = \begin{cases} 0, x \leq 1.004 \\ \frac{x-1.004}{0.086}, 1.004 \leq x \leq 1.09 \\ 1, 1.09 \leq x \leq 1.3 \\ 0, 1.3 \leq x \end{cases} \quad (A.49)$$

The Output Nocturnal Blood Pressure Profile variable has the linguistic values *"Extreme Dipper"*, *"Dipper"*, *"Non-Dipper"* and *"Riser"*, for each upper and lower membership functions. The values of each point of the fuzzy system are presented as follows.

$$\mu extdipper(x) = \begin{cases} 0, x \leq 0.4 \\ 1, 0.4 \leq x \leq 0.6655 \\ \frac{0.8-x}{0.1345}, 0.6655 \leq x \leq 0.8 \\ 0, 0.8 \leq x \end{cases} \quad (A.50)$$

$$\mu dipper(x) = \begin{cases} 0, x \leq 0.787 \\ \frac{x-0.787}{0.024}, 0.787 \leq x \leq 0.811 \\ 1, 0.811 \leq x \leq 0.889 \\ \frac{0.9102-x}{0.0212}, 0.889 \leq x \leq 0.9102 \\ 0, 0.9102 \leq x \end{cases} \quad (A.51)$$

$$\mu nondipper(x) = \begin{cases} 0, x \leq 0.898 \\ \frac{x-0.898}{0.025}, 0.898 \leq x \leq 0.923 \\ 1, 0.923 \leq x \leq 0.9821 \\ \frac{1.02-x}{0.0379}, 0.9821 \leq x \leq 1.02 \\ 0, 1.02 \leq x \end{cases} \qquad (A.52)$$

$$\mu riser(x) = \begin{cases} 0, x \leq 1.006 \\ \frac{x-1.006}{0.084}, 1.006 \leq x \leq 1.09 \\ 1, 1.09 \leq x \leq 1.3 \\ 0, 1.3 \leq x \end{cases} \qquad (A.53)$$

Appendix B
Graphical User Interface

To visualize the results of all the soft computing techniques used in this research, a graphical user interface is designed, which can be seen in Fig. B.1 and each of its parts are described below:

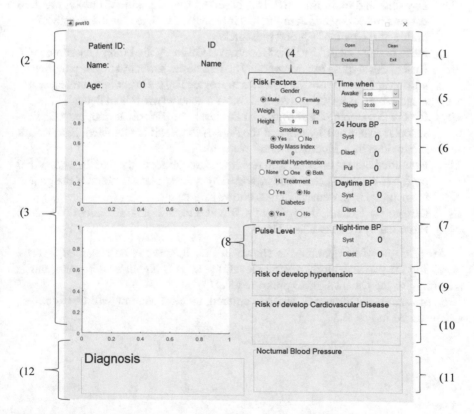

Fig. B.1 Graphical user interface

© The Author(s), under exclusive license to Springer Nature Switzerland AG 2022
P. Melin et al., *Nature-inspired Optimization of Type-2 Fuzzy Neural Hybrid Models for Classification in Medical Diagnosis*, SpringerBriefs in Computational Intelligence, https://doi.org/10.1007/978-3-030-82219-4

1. In the Upper Right Part There Are Different Buttons for Its Operation:

    (a) Open: It is used to find the file where the patient's blood pressure records are located.
    (b) Evaluate: Patient readings are evaluated by the neural networks and fuzzy classifiers used.
    (c) Clean: To clean the entered records.
    (d) Exit: Exit the interface.

2. Patient data is displayed, such as patient number, name, age.
3. The records of the patient's blood pressure obtained throughout the study are displayed graphically.
4. Risk factors: The data corresponding to the patient's risk factors are entered. These will be the input to the artificial neural networks to obtain the risk of developing hypertension and the risk of developing a cardiovascular event.
5. Time when: The time in which the patient falls asleep and wakes up is entered.
6. 24 Hours BP: It is the Trend of Blood Pressure Obtained by the Modular Neural Network
7. Day time and night time BP: It is observed how the patient's blood pressure behaves in the day and at night, this information will help us to know the night profile of the patient's blood pressure.
8. Pulse Level: Obtain the Heart Rate Level Given by the Fuzzy Classifier
9. Risk of developing hypertension: The risk of the patient in developing hypertension in 4 years is obtained, which is provided by the artificial neural network, if the patient is hypertensive, it shows a legend where it is indicated.
10. Risk of Developing Cardiovascular Disease: The Risk of Developing a Cardiovascular Event in 10 years and the Age of the Heart is Obtained. Said Result is Provided by a Modular Neural Network.
11. Nocturnal blood pressure: The result corresponding to the patient's nocturnal profile is displayed, which is provided by a fuzzy classifier, and if the patient is nocturnal hypertensive, it also shows said result.
12. Diagnosis: Presents the patient's Blood Pressure Classification, Which is Provided by a Fuzzy Classifier.

An example of its operation is shown in Fig. B.2. It can be observed that, by providing the patient's blood pressure level, the label is highlighted in color, this is according to the patient's blood pressure level.

In the same way, if no risk factor is entered, an alert window will be displayed, as illustrated in Fig. B.3.

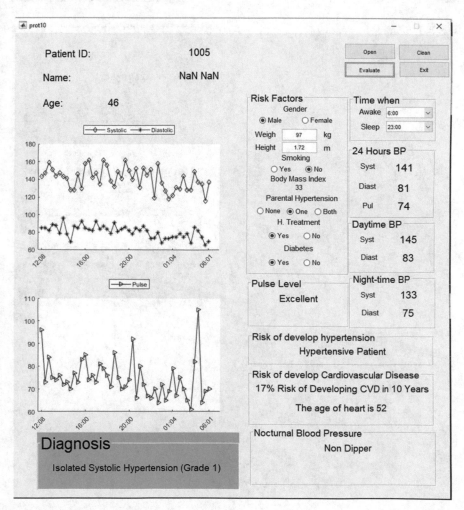

Fig. B.2 Results obtained by GUI

Fig. B.3 Warning messages

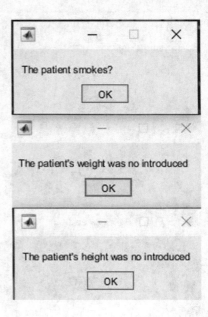

Index

Printed in the United States
by Baker & Taylor Publisher Services